PACKAGING STRATEGY

Packaging Strategy

Winning the Consumer

EDITED BY

MONA DOYLE, CMC

Publisher, *The Shopper Report*
President, Consumer Research Network, Inc.

TECHNOMIC
PUBLISHING CO., INC.

LANCASTER · BASEL

Packaging Strategy

a TECHNOMIC® publication

Published in the Western Hemisphere by
Technomic Publishing Company, Inc.
851 New Holland Avenue, Box 3535
Lancaster, Pennsylvania 17604 U.S.A.

Distributed in the Rest of the World by
Technomic Publishing AG
Missionsstrasse 44
CH-4055 Basel, Switzerland

Printed in the United States of America
10   9   8   7   6   5   4   3   2   1

Main entry under title:
  Packaging Strategy: Winning the Consumer

A Technomic Publishing Company book
Bibliography: p.

Library of Congress Catalog Card No. 96-60500
ISBN No. 1-56676-298-7

TABLE OF CONTENTS

FOREWORD

The best packages and labels make winners of consumers by enabling them to readily make the choices that best fit their needs. As packages and labels become easier to use, they enable consumers to make better decisions and to be more discerning. As consumers become more discerning, they demand better information and better performance from their packaging and labeling.

During my years as consumer advisor to Presidents Kennedy and Carter and Giant Food, I've seen dramatic improvements in packaging and labeling and in consumers' use of information. Of course, there is still room for improvement in many areas. With the surge in the aging population, we need more senior-friendly packaging and labeling. We need more rational serving sizes on food packages. We need to make the awkward move to metric sizes. We need to keep improving freshness dating or time/temperature indications of product integrity.

I'm very pleased about the responsiveness of packaging and labeling to consumer needs—and delighted to see the title of this book—and its implication that winning the consumer is ultimately a win for the producer and retailer as well. My congratulations to Mona, the authors she has assembled, and the readers who are willing to address the challenge of winning over the consumers of tomorrow.

ESTHER PETERSON

PREFACE

Pressures on packaging are intensifying. Packaging will have to earn its keep and keep its edge through this period of dramatic change in the retail marketplace and the economy.

Packagers and marketers will have to develop transitional as well as long-term strategies to respond to the emerging demands of electronic retailing; the need to perform in both physical and electronic environments; and the time, convenience, and value demands of increasingly sophisticated consumers who increasingly expect performance, value, and social responsibility from packages, as well as from the products they contain.

I hope that the kaleidoscope of packaging views and recommendations presented in this book will help readers find ways to meet these demands.

MONA DOYLE, CMC

ACKNOWLEDGEMENTS

My thanks, first, to the thousands of Consumer Network panelists who have shared their packaging perceptions, compliments, and frustrations. Until very recently, when they have begun to think of packaging as helpful and responsive, most of them expressed wonder about why those who design and make packages don't care about those who use them. They *knew* that the makers didn't care because they have broken a nail or chipped a tooth trying to open a tough package; given up when arthritic hands couldn't get the grip they needed; or gritted their teeth on labels they couldn't understand, serving sizes that didn't make sense, or instructions that seemed to be meant for someone else. Today, they are increasingly quick to praise and purchase a better package.

My thanks also:

- to authors Lorna Opatow, Lynn Scarlett, Phil Fitzell, Greg Erickson, Tony Adams, Gerald (Jerry) P. Meier, Kristen McNutt, Sheldon Sosna, Chuck Mittlestadt, and Herb Meyers, the experts who accepted my vision of a multiperspective approach and contributed their thinking and writing time to this book
- to Campbell Soup for sponsoring my *Packaging at the Crossroads* seminar
- to Procter and Gamble's Corporate Packaging Group, who took The Consumer Network's assessment of their packaging to the drawing boards—and responded to almost every criticism we delivered
- to Westvaco's Ovenable Packaging Group, who gave us the

packaging research assignment from which we learned how to recognize a winner

- to Waldbaum's, Winn Dixie, and Certified Grocers, who co-sponsored a Senior Shopping Study that enabled us to see that, when aging is cumbersome, packaging that is easy to handle makes life a little easier
- to Confab, producer of private label sanitary products, for the opportunity to assess their products and packages with consumers of all ages, from infancy through pregnancy to the fast-growing and increasingly active seniors who want to be liberated from bulky and hard-to-use packaging
- to the Food Marketing Institute's Consumer Affairs Council, which gave me the opportunity to chair a Packaging and Labeling Task Force, and most especially, to Karen Brown, Mary Ellen Burris, Kathy Lapier, Odonna Matthews, Kit Searight, Katherine Smith, and MaryLu Waddell, the supermarket women who helped me to blend consumer affairs into market research and became my friends as well as colleagues
- to Ben Miyares and Sophia Dilberakis, who encouraged and published my Consumer Reaction columns in *Food and Drug Packaging Magazine* for over ten years
- to consulting colleagues Robert Kahn, CMC, Bill Altier, CMC, and Art Kranzley who have been mentors and friends through this and many other efforts
- to consulting associate Joe Walker, who successfully challenged my thinking—as usual
- and to Esther Peterson, who taught me how to hear the straws in the wind

Introduction

The Consumer Growth of Packaging Power

Packaging power is growing. It's growing because packaging is delivering significant consumer benefits and competitive advantages at the same time as advertising is losing much of its traditional reach and clout. It's growing because packaging is successfully finding ways to meet consumers' conflicting goals of more convenience *and* less packaging. The selling power of packaging that is relevant to consumer needs is gaining brand marketing respect while environmental pressures dictate that packages practically disappear, or at least that they become lighter and smaller or leaner and meaner. Consumers are responding to packaging excellence with increased awareness and purchase dollars. Packaging is not only selling products in the store, but reselling them during their use at home, at work, or in the car. The selling power becomes a self-repeating cycle: because of packaging's increasing effectiveness as a sales tool, marketers are increasingly turning to packaging as well as advertising and promotion to keep their products moving.

The decline of mass advertising as the primary driver of packaged goods sales has placed more of the total communication burden on packaging. Consumers' pressing needs for saving time and reducing the hassles of everyday life put further demands on packaging. Easy to choose and use packages for shredded cheese, for example, turned a novelty item into a major product category. Stunning designs gave many store brands a new stature. Trigger sprays changed the dynamics of household cleaners. New labels enabled food shoppers to make quick decisions on which products met their dietary needs. Wide-mouth bottles for juices reshaped the whole beverage category. Unit-dose packaging increased compliance with drug regimens and now accounts for a growing percentage of the sales of both prescription and OTC drugs.

An avalanche of new products has changed the retail landscape and intensifies competition for retail shelf space. It also makes packaging and shelf presence more important than ever. Consumers who participate in focus groups have more to say about their packaging expectations. Increasingly sophisticated, they can be observed in the supermarket, drug or discount store seeking information as well as function from the package. As their choices proliferate and value perceptions develop, the positioning role of the package becomes more critical in defining the product. All in all, the power of packaging to influence purchasing decisions has increased during the 1990s and will continue to do so unless (1) advertising regains its dominance to become consumers' primary source of trusted information or (2) packagers lose sight of consumers. To hold its power, packaging must serve as a source of added value, keep up with consumers' needs, and continue to be a primary source of product information.

The Corporate Growth of Packaging Power

Demand for chain store shelf space is huge, especially for packaged products. Competition for that space is ferocious. The spiraling costs of getting the space can be staggering, and getting it is only the beginning. Generating the sales required to stay on the retail shelves is as costly as getting there in the first place.

With mergers, space demands and costs all at all-time highs, with competition for shelf space more intense than ever, and with the risks and costs of failure high enough to make even those with deep pockets wince, marketers who compete for shelf space are asking more of packaging than ever before. With 25,000 new products a year crowding both old favorites and less than blockbuster new items off the shelves, the demand on packaging to contribute to both purchase and repurchase decisions is escalating. Many ambitious marketers know that a blockbuster package can double the odds of success. But in the rush to bring new products to market, they rarely have the time, patience, or insight to make it happen.

Advertising can no longer be counted on to create enough demand to pull products from the shelf. It has begun to depend on package images for fast communication. Although advertising, media, and retailers come together in delivering packaged consumables, advertising is increasingly fragmented. Consumers, en masse, are hard, next-to-impossible to reach. Marketers increasingly rely on promotions to support advertising, but packages that get as far as the chain store shelf reach consumers 100 times more often than any ad or promotion. And more than advertising or promotion—more than anything but word of mouth or personal experience—packages are the primary communicator. Usually, they are the sole intermediary between the product and the consumer.

Packages that win both on and off the shelf have enormous power. They accelerate the first purchase decision. They shape the consumers' experience with product use. They influence attitudes and decisions about repurchase. Recognizing this power and struggling with intensifying competition, packaged product marketers have begun to look to the package as a central, rather than peripheral, part of the marketing effort.

As a result, package improvements are becoming much more common, but they have a long way to go to catch up with other kinds of product improvements and advertising changes. Structural package benefits such as "easier-to-open" and "now resealable" are beginning to be touted on front panels, the precious face of the package that has long been known as the selling area.

McDonald's changeover to new packaging that looks eco-friendly was independent of product changes. And it took place with little fanfare. Consumers who noticed the packaging change thought the new bags, cups and trays were "nice." One consumer told us the new packaging made the food seem a little healthier. The financial community barely noticed the change. No one was terribly excited!

In sharp contrast to McDonald's extensive but quiet packaging change, the new product machine has been turning out and introducing new products at dramatically increasing rates throughout the last two decades, and almost always tries to generate excitement, which, in turn, generates product trial. In 1995, more than 25,000 new products were introduced. In spite of heightened package awareness, only a tiny proportion of the new products were introduced in new packages. Minor source reduction modifications were widespread, but "functionally apparent" packaging innovations with benefits that are obvious to consumers were few and far between although good packaging—especially when it makes functional sense to consumers—contributes greatly to a new product's success.

Poor and nonfunctionally-transparent packaging that doesn't communicate its value contributes to the failure of the rest. A hefty majority of new products fail, sometimes because their packages fail to meet consumers' expectations. Mediocre packaging rarely gets blamed for the failures, just as great packaging rarely gets the credit for success outside of the technical packaging community. Many that succeed have packages with obvious benefits that fit the product, meet or exceed consumers' expectations, and clearly do the job that consumers believe needs to be done.

The Time Barrier

The persuasive power of advertising people isn't the only reason that packaging has been slow to get all the recognition it warrants. Time works against packaging changes too. New product development has been the lifeblood of the packaged goods business and timesaving has become increasingly important in the new product development process. More and more emphasis has been placed on shortening the time from idea generation to appearance on the shelf. In spite of dramatic developments in computer-aided design, new packaging takes longer to develop than new advertising or new product formulations, especially if new product machinery is required. It follows that most new packaged products are launched in me-too packages.

Packaging and the Marketing Team

Packaging has won a seat at the marketing decision table. Companies are beginning to recognize the value of integrating packaging into their marketing teams, and marketers and sellers are paying more attention to packaging as its selling power becomes clear to them. Key executives at major companies are beginning to talk about brand leveraging through packaging. The American Marketing Association has staged its first conference about marketing success through packaging. Some marketers are even beginning to look at the whole cycle of consumers' relationship to their packages and to how well their packages work for the consumer from prepurchase consideration to eventual disposal.

Some companies are beginning to integrate their marketing efforts to include all the layers of communication and decision making involved before their products reach the consumer. Other companies are grouping their marketing communications and sales promotions around products in the ways that consumers perceive them. Even Procter & Gamble, the quintessential brand competitor, has shifted from brand-centered to category-centered management.

Packaging continues to gain recognition as a marketing tool because more and more spotlights are directed at today's packages. For example, supermarkets are increasingly relying on color images of actual packages in their ads. Major advertisers are getting ready for a future of interactivity when consumers will have access to advertising and packaging details from their cars and "teleputers." Consumers are already doing more and more of their food purchasing (and consuming) on the basis of packaging that facilitates grab-and-go purchasing and in-car eating. Today's supermarkets have package libraries on hand in their computerized advertising departments. Almost in real time, library color photographs of packages are dropped into the ads, so that shoppers see the virtual package in their supermarkets' advertising or catalogs whether they are looking at a printed ad or circular or shopping via cyberspace. With so many critical roles to play, packaging will eventually become a full marketing team member.

Options and Opportunities

There are many strategic reasons to change packaging, so many that it is ironic that the number of changes taking place doesn't approach the number of product changes. The strategic reasons for changing an existing package or developing a new package include:

- achieving legal compliance
- accommodating a size change
- adding to product and brand value
- attracting shoppers' attention
- countering bad publicity
- creating good will
- demonstrating social responsibility
- improving value perception
- increasing consumer satisfaction
- improving safety perceptions or meeting safety standards
- adapting to the state of the art
- reducing pilferage
- responding to consumer frustration and complaints
- responding to changing consumer needs
- responding to changing demographics
- responding to competitive pressure
- revitalizing a brand
- combinations of above

Inferring Strategies

Some companies and industries have been open about their packaging goals and strategies. P&G, for example, has gone public with its source-reduction goals and has taken leadership positions in bringing source-reduced packages to the marketplace, going so far as to state the amount of source reduction achieved on the front panel of some of their packages.

Other companies have been less forthcoming. But you don't need inside information to infer key parts of the strategy that underlies many new packages.

Packaging change: Pegboard candy packs introduced to replace and compete with bars.

Inferred strategy: Increase shelf impact and reduce pilferage.

Packaging change: Plastic canisters with reclosable measuring-cup lids and finger-grip waist moldings for Tang, Kool-Aid and Country Time powdered drinks.

Inferred strategy: Add brand value by making powdered drinks more convenient to use and hold at a time when no-effort, ready-to-drink beverages are gaining market share. The easy-hold feature is probably designed to reduce kid spills and improve value perceptions.

Packaging change: Lifesaver has increased the number of mints and total size of its traditional packs without raising the price.

Inferred strategy: Exceed consumer expectations and build value perception.

Packaging change: Plastic 9.5 lb refill packages for P&G powdered laundry detergents are designed to pour into boxes or plastic containers sold by mail order.

Inferred strategy: Respond to consumer interest in refills and large, value sizes and continue to enhance reputation for environmentally responsive packaging innovation.

Packaging change: Pepsi introduced a 20-ounce, narrow-waisted plastic bottle.

Inferred strategy: Compete head on with Coke's nostalgia bottle.

Packaging change: Breyers introduced a rectangular canister package to replace the troublesome one-piece rectangular carton that has frustrated nonpremium-priced home ice cream users for years.

Inferred strategy: Add value to a well-loved, middle-priced ice cream brand that was being squeezed from above by the premium-priced resealable rounds and from below by the value-priced store brands.

Packaging change: Campbell's departed from its icon red and white soup labels to incorporate a picture of a bowl of the named soup on each label.

Inferred strategy: Respond to rising proportion of shoppers with low levels of English literacy, speed up shopper's search time for preferred flavors and revitalize brand.

Packaging change: "Feel the curves" is what Coca-Cola ads and posters are saying to consumers about its shapely plastic bottles.

Inferred strategy: It's fun! An icon! Tradition! It's a double entendre that consumers have no trouble understanding and enjoying as much as they enjoy the drink inside. It's cool. It's hip. It's fun for adults. It's an intimate message for an intimate experience with a very familiar product. More than that, it's almost a recognition of the consumer-packaging relationship—a relationship that is accompanied by frustration and guilt, familiarity and irritation, satisfaction and fatigue, and sometimes a great deal of pleasure. It's retro when retro is the rage. It's nostalgia that impacts as many as four generations, speaking not only to the X-generation, but also to today's high school students, as well as their parents and grandparents. It isn't "the real thing" (the old glass bottle), but it's awfully close. In fact, it's a 1990s version of the real thing—a compromise for the kind of plastic convenience that makes sense for today's lifestyle. It can't have the same icecold mouth-feel as nostalgia associates with the thick-lipped glass Coke bottle. "Nothing could ever replace the glass Coke bottle, but it's great." "It makes you want to have one, even if you only drink mineral water now." It's easy to recognize, easy to drink from while on the go. It's easier and more pleasant to hold than the can. It's easy to open and reseal, easy to store, and easy to drink with one hand while walking or driving. It stands in most (not all) car cup and can holders, sits comfortably on the shelves of cabinets or refrigerator doors, and fits easily in a briefcase, purse or backpack.

Unlike New Coke or Clear Pepsi, both of which created high interest before they fizzled, the new Coke bottle is readily understood and makes a positive difference in perceptions. It differentiates Coca-Cola from other beverages and other brands. And it isn't something that is likely to be quickly duplicated by store brands.

New packaging rarely creates this much interest in an established product. But it does happen. Sometimes new packaging goes beyond a single brand to create a new style or even a new category.

Packaging change: Tide and Cheer refills make the statement "80% LESS PACKAGING THAN CARTONS" right on their front panel, just under the product name. And they don't stop there: They feature 25% post-consumer recycled content and recyclability.

Inferred strategy: The plastic refill packages for P&G powdered laundry detergents respond directly to consumers' perception that there should be more refills on the shelf. Less packaging is perceived as an improvement—as long as it doesn't mean less product. The refills were widely nominated as the best package of the year 1994. They are admired for their bold-source reduction sell copy, which consumers find at least 99.75% credible and very educational to boot. Almost 90% of the shoppers The Consumer Network surveyed in 1994 agreed that "the less packaging, the better." The strongest agreement came from females between 30 and 49; the least from males, who are less concerned about packaging problems in general. (Males are more likely to have the hand strength, or to carry pocket knives, that facilitate opening packages without breaking fingernails. They are also less likely to see packaging as an environmental villain.)

Ongoing Opportunities

Gable-top paperboard containers are vulnerable to a better package because their performance is far short of excellent. They are still around because they leak less; because some of them appear with twist-off closures that look funny but solve the opening/reclosing problem; and because no one has come up with a cost-effective alternative—at least not yet.

Paperboard cereal boxes with plastic liners are still around because they are cost effective and provide unbeatable selling space. (Prediction: They won't survive the 20th century without further modification.)

Large paperboard boxes of detergent with their infamous "Push Here To Open" closures are slowly but surely losing share to liquids packaged in plastic.

Confusing Packaging Can Limit Category Growth

Just as a real improvement can create a category, failure to develop the right package contributes to the failure of a new category. To be effective in stimulating trial and repeat purchase, the points of difference of a new product or a new package must be clear to the consumer. A new barrier film that improves product shelf life may be of great value and deliver a better-quality product to consumers over a longer period of time, but the improvement doesn't speak to consumers' frustrations or wishes and, therefore, will not stimulate an increase in trial. A beautiful package that doesn't communicate the benefits, definition and position of a "new" product carries part of the responsibility for the product's failure. At the time of this writing, packaged refrigerated entrees and dinners have failed to win a major franchise in the United States in part because none of the packages in which these products have appeared has been functionally transparent to consumers. The packages do not communicate that the products inside are different from or better than frozen food. In fact, some consumers who tried packaged refrigerated products referred to them as "these frozen foods," or simply "frozen foods," because the new packages failed to clearly differentiate them from frozens in consumers' minds.

It is possible that the double reminders boldly stated on new Nestle's products will make the difference: The new products change the Stouffer name to Stouffer Refrigerated and add the brand "Take Home Entrees For Two"—strongly clarifying the idea that this is something very different from Stouffer's frozen red box dinners. Certainly, the Stouffer effort has a far better chance of being immediately understood by consumers than any refrigerated entree we have seen before.

Some New Packages Create New Products

Sometimes it works the other way, however. Sometimes the package is better liked than the product it contains. The Pringles canister is an example of a great package waiting for a great product. Pringles never captured a major share of the U.S. snack market, but the product and its unique package have been a snack presence for many years. New flavors, formulations, ads and a great package have combined to keep Pringles alive. Although the product has improved over the years since its introduction, the great package continues to be its reason for being.

Some innovative packages are major factors in new product introductions.

- packaging category creation 1: The bottled waters that consumers carry to control their weight and quench their thirst is a category created by convenient and attractive packaging.
- packaging category creation 2: Budget-priced frozen entrees from Budget Gourmet and Michelina won consumer credibility for their strong value positioning through a packaging tray that communicates "minimal packaging, maximum food value."
- packaging category creation 3: L'Eggs famous, and then environmentally infamous, eggs are no longer on the supermarket shelves, but they created a brand that lives on after the egg-package has been replaced with a less costly and more environmentally sensitive package.
- packaging category creation 4: Le Menu frozen dinners were introduced on a plastic-domed round plate with added layers of protection in the form of a foil seal and a paperboard carton. Le Menu changed the perception of frozen dinners for many young consumers in the early 1980s. A young man in a packaging focus group said that one experience with Le Menu showed him that he would indeed survive his divorce and the lack of a wife to fix his meals. Changing consumer expectations forced some

dramatic changes in the Le Menu packaging, but the brand is still in the stores—and it's still packaged on a round plate.

Reasons for Resisting Packaging Change

In his chapter on "Package Power," Chuck Mittelstadt makes the point that "contracts" exist between loyal consumers and brand marketers which "provide" that no significant changes in product, package or pricing can be made without their consent. When such a contract is broken, consumers feel justified in leaving the brand and even feel foolish or "had" if they continue to purchase it. Tacit recognition of this contract is one of the barriers to packaging change. Some marketers believe that loyal users see any packaging change as a breech of that contract. Indeed, many consumers believe that most "improvements" to products and packages are primarily excuses to raise prices.

Even though a strong case can often be made for an ambitious packaging change, the contract between loyal consumers and brands is only one of many solid reasons to hang tough. Here are a few more:

- capital cost of packaging machinery
- financial analysis showing low or uncertain ROI
- commodity cost of alternative packaging materials
- time required to develop package improvements
- habit
- market share being maintained
- "it's not broken, don't fix it" mentality
- belief that consumer franchise is built on existing package
- production- rather than market-driven culture
- reluctance to yield traditional selling benefits, especially billboarding
- perceptions of cost increases due to change or improvements
- available improvements not good enough
- consumer or competitive pressure not strong enough
- disappointing experience with earlier packaging changes

Cereal is the preeminent example of strategic resistance to consumer demand for packaging change. The rigid billboard packaging by all the major players in the RTE cereal industry hangs on after years of consumer frustration. Many academic, regulatory and consumer group studies have blamed cereal industry concentration on high cereal prices. So far, none have focused their criticisms on packaging, but resistance to change in this category would make an exciting Harvard case study.

The primary RTE cereal packaging question is as basic as: "To change or not to change?" On the pro side of change, better packaging would expand cereal's snack and grab-and-go usage, both of which are discouraged by the tall and tippy box and liner. Better packaging would be more consumer friendly, easier to open and reclose, and more in sync with 1990s expectations. Better packaging might even expand purchasing in humid and coastal areas, especially where insect problems are widespread and many consumers feel it is necessary to refrigerate packages after opening. The con side of change includes the growth of cereal usage in spite of the unfriendly package; the perception that cereal prices are too high even without any improvements that might add cost; and the powerful sell-space billboard of the traditional package.

When cereal packages do change for the better, they will probably go like dominos. This writer suspects that at least one big change will take place well before the turn of the century. The question is how, and when, and in just what kind of package providing what assortment of benefits?

Some would bet on flexible plastics, perhaps stand-up pouches a la Lipton Noodles Plus with gusseted bottoms. Others bet on a rigid paperboard package with a plasticized liner that keeps out moisture and bugs and resembles the new Breyers ice cream package—a square canister made from paperboard that is easy to open and relatively easy to reclose. The shape of the new package might resemble the stubby Shredded Wheat package or an Ultra detergent package. Perhaps the new cereal package will even be called the Ultra package, just as compressed detergents in stubby reclosable packages were called "Ultra."

A strategically viable cereal package may or may not yet exist. Consumer dissatisfaction with the present package will continue to exist until it is resolved. The pitch of the dissatisfaction has eased off in recent years as frustration with cereal prices put packaging frustrations on a back burner. Consumers would be less willing to go on fighting with the cereal box if they weren't reeling from high cereal prices or feeling that there is less cereal in the box. Compared to the high level of price frustration, the fact that the package isn't up to speed becomes trivial. Packaging improvements might even exacerbate high price perceptions among consumers who don't share the feeling that the "awful packages make the awful prices even worse."

The Cross Section of Contributors

In recognition of packaging's growing importance to the marketplace as a whole, this book presents an exceptional cross section of perspectives and viewpoints. The views of expert package designers, trade journalists and researchers are echoed and challenged by consumer, advertising, supermarketing, and packaging machinery producer perspectives. The breadth of the contributor cross section is intended to broaden the perspectives of packagers and marketers as they develop packaging strategies for a fiercely competitive marketplace and increasingly sophisticated consumers.

MONA DOYLE

Mona Doyle is a consumer perceptions expert specializing in strategic marketing, packaging, and consumer relations. She is founder of The Consumer Network, Inc. and publisher of *The Shopper Report*, a consumer-relations monitor for marketers.

Her firm helps companies research marketing and packaging strategies and programs by means of mail surveys, focus groups, mystery shopping and competitive analysis.

Mona Doyle is a Certified Management Consultant (CMC) of the Institute of Management Consultants. She serves on several boards; speaks to business groups around the world; has appeared on Face the Nation, Good Morning America, and National Public Radio; and is frequently quoted in the *Wall Street Journal, Business Week, Family Circle, Advertising Age, Supermarket News, The New York Times, The Chicago Tribune,* and *The Philadelphia Inquirer*.

An honors graduate of the University of Pennsylvania in psychology, Mona has done graduate work at Temple, Drexel and Cornell universities. She is an avid walker and tennis player, an involved mother and grandmother, and a resident of Voorhees, New Jersey, and Philadelphia where The Consumer Research Network is based.

Getting It Right:
Research to Ensure
Successful Packaging

While it doesn't always get the respect it deserves, packaging is exceptionally powerful!

No one buys an empty package, and we can rarely buy a product without a package. This interrelationship is key to understanding the power of packaging to influence perceptions of, and reactions to, the brand. Unless there is something wrong with the package, consumers tend to look through it to the product. As a result, their ideas about the brand are refracted by the package.

Companies understand this, and it is one of the driving forces behind the use of marketing research for package development.

At the start of the overall package planning process, a set of criteria or objectives will be established. These are often amended as new information is received. Some of the objectives are exclusive to the package, others are shared with the business and marketing plans. In any case, the packaging, business and marketing objectives need to be compatible.

Packaging objectives usually include an overall statement of physical and graphic packaging goals, followed by specific objectives. For example, the overall goals for a paint container were that it be "dripless" but resealable. Graphics were to be designed to register the brand name, differentiate colors and types of paint in the line, and reinforce the convenience of dripless painting.

The specific objectives then focused on meeting distributor and retailer requirements, appealing to a specific group of target customers, etc.

Here are some goals that are common to most packages:

1. In-home
   • be functional (easy to open, close, store, use)

  - reinforce product satisfaction
  - remind customers when to repurchase
  - remind customers to buy the same brand
2. In-store
  - gain attention on the shelf
  - identify the product
  - identify the brand and differentiate it from competition
  - say something good about the product
  - induce consumers to buy
3. General
  - protect contents
  - satisfy distributor requirements
  - satisfy retailer requirements
  - satisfy legal requirements
  - be adaptable to illustration in both color and black and white

These criteria appear to be reasonable, but they assume the availability of a considerable body of knowledge.

Much of the early research related to new product development can also be used for the package. Whether deliberately planned to do so, research that helps define the target market, establish brand positioning, identify the competition, evaluate systems of distribution and determine those factors that help generate favorable reactions to the product can also be used to formulate the criteria that guide package development.

Also, results of ongoing studies for established brands may suggest the need for a packaging change. For successful brands, research may be undertaken specifically to determine whether the package should be changed and, if so, to provide information to guide the redesign.

A company or brand's image is usually a critical consideration in formulating packaging goals. Surveys to measure reputation or image can show unsuspected differences in competitive strengths and weaknesses. The favorable differences can be used to establish an effective positioning: one that combines a group of characteristics to create uniqueness compared with other brands.

The position may be based on a particular product characteristic: "fast relief" for a remedy, "energy efficient" for a ceiling fan, "sun-resistant" for paint, etc. Or the position may focus on how the product is used: "heat and serve in the container" foods, "comb-dispenser" hair colorant. Positions may also be based on methods of product delivery, container size, product price, user characteristics, or brand image.

Effective positioning requires not just uniqueness, but a superiority that is relevant to intended customers. In order to accomplish this goal, marketers need to identify target customers and understand how and why they use the product or service. During this investigation, it will become apparent that it is impossible to be all things to all people.

Many companies identify their target customers so broadly that there is nothing to work with. For example, here is a target description which a food processor initially gave to both its advertising agency and package developers:

- young homemakers 18–34
- middle- to upper-middle income groups
- presently use other convenience foods

This kind of description does not provide direction, nor is it helpful in establishing goals for a packaging or promotional program. In fact, it creates a special problem if surveys are used to judge the effectiveness of the program, because the interviews may not be conducted with the right people—potential users of the brand. If so, the information obtained will be misleading, no matter how carefully the survey is conducted.

This particular company already had a great deal of information. Several surveys had been conducted during the product development stage and other marketing information had been collected from public sources. The material was there but had not been coordinated or analyzed to identify the target market. As a result, a Data Review Analysis study was undertaken. It included information from government departments, associations, trade and consumer media, as well as syndicated and other surveys, but did not require carrying out a new survey.

Valuable information came from this Data Review. For example, it gave direction for revising the target market description. Compare this final description with the one given above:

- men *and* women who believe they are pressed for time
- people living alone or in small families
- users of specific types of convenience foods
- willing to pay more for taste *plus* convenience, but not for convenience alone
- regularly shop certain sections of the supermarket

An accurate description of the target customer group is critical to developing effective packaging.

Packaging development is a sophisticated balancing act geared to satisfying a series of goals or criteria that are often mutually conflicting. Developers need to keep the goals in mind throughout the package development process, while retaining a clear idea of the importance of each.

Initially, research is used to help guide the formulation of packaging objectives. When packages are ready, research is used to measure their effectiveness; the degree to which goals have been met. The research focuses on four broad areas: *functional, visual* and *communications effectiveness, and esthetic appeal.*

Functional effectiveness has to do with the performance of the physical package, including legal requirements related to child-resistant and tamper-evident features. For consumers, measures of functional effectiveness may involve both handling characteristics (ease of opening, closing, storing, using) and perceptions of freshness and cleanliness (based on their opinions of protective features).

Ease of handling is usually measured based on observations of the product/package in use (if actual packages are available) or observations of pretended use (based on models). At some point in testing new products, different types of physical packages may be tested along with the product.

There are two circumstances where it is especially important to include a test of the physical package along with the product. The first is when a less expensive packaging component is being considered for an established product. In this case, it is important to be sure there is no corresponding decrease in product acceptability. The second case is where the physical package has been used to create a new product form as was the case with Soft Soap, aerosol dessert toppings and spray bandages.

If the physical package represents a technological breakthrough such as aseptic containers, there may be several special research problems. In the early development stages, few filled packages will be available and those that are will not meet final performance standards. Also, lack of an adequate verbal description may hamper research to determine how acceptable the package is for selected purposes.

These problems can usually be solved through small-scale studies using qualitative research techniques. For example, ideas about functional effectiveness can be obtained based on sketches, mock-ups, prototypes and test-run samples. Consumers can be interviewed in groups or individually using in-depth or standard interviews at central location facilities, at home or at work. Or they may be asked questions about the package during a product test.

Questions about functional effectiveness can be asked directly. Consumers are the best judge of whether they find the package easy to pour from or hard to open, and they have their own ideas about whether a package will keep the product fresh.

Unlike functional effectiveness, measurements of a package's *visual impact* are more difficult to obtain. This is partly because of the complexity of what we call "visibility." Generally, there are three types:

1. *In-store stopping power based on a shelf presence that is established over time:* This occurs because the majority of shoppers in the aisle where a product is displayed are not buying the category that day. Over time, as they repeatedly pass the shelves or become aware of advertising, or both, they recognize that there is something new or different. At that point, they may consciously notice the brand when they shop the category. In general, we cannot measure the ability of a package to establish a shelf presence over time.

2. *Recognition of current packages:* This is based on equity built through advertising, familiarity and use. That is, people will recognize the brand based on overall shape, or color, or one or more general graphic configurations without really "seeing" the package or reading the brand name.

3. *Visibility tests:* For current packaging, visibility tests measure the recognition resulting from familiarity. For new products and redesigned established brands, visibility tests are measuring legibility and, to some degree, attention value. Shelf impact studies are complicated by the fact that vision is controlled by the mind, rather than the physical structure of the eye. Optical illusions are based on these deficiencies in the physical mechanism of seeing. For example, the length of two lines are identical, but one line may seem to be longer than the other because of the treatment of the ends.

A package's noticeability can be increased by placing it at eye level on the shelf, by increasing its number of facings, by surrounding it with packages in contrasting colors or of contrasting brightness, by in-store shelf markers and similar devices, and by increasing advertising. These factors all need to be taken into consideration in planning shelf-impact studies.

Regardless of the test method used, standards of acceptability for shelf-impact tests should be established in advance. Standards may be set in relation to competition and varied depending on whether current or potential customers are being interviewed. They will take into account the fact that familiarity biases favor a current package because the known tends to be noticed before the unknown.

The most commonly used methods for testing shelf visibility involve using an actual or virtual simulated shopping visit, a tachistoscope (T-Scope) or an eye-movement test involving a single package, a competitive array or a mock shelf display. The T-Scope, eye-movement and computer-simulation tests rely on pictures that may distort some colors, and that tend to create a more orderly display than would be found in a store.

In a simulated shopping test, a rough approximation of a shopping environment is reproduced on a computer screen or in a central interviewing facility. Products are displayed within a section of shelves and people "shop" as they would if in a store. They view products while actually or apparently in motion, looking at what interests them and ignoring what does not. The setting helps create some of the visual confusion that occurs in a store.

In a T-Scope test, a device attached to a projector controls the length of time a picture is shown on a screen. Exposures are controlled for increasing amounts of time. After each exposure, people are asked about what they saw. Results are evaluated based on speed of, and total recognition. T-Scope results rely on what people say they saw. While people often mention colors, the technique tends to emphasize brand name readability for new products or redesigns and recognition for current products.

In an eye-movement test, the respondent controls the length of time the picture is viewed. A device records where the eye moves and where it stops over the surface of the picture. Results are evaluated based on this record.

Regardless of technique, the results of visibility tests are not usually used as the sole basis for accepting or rejecting a package. Packages must also meet communications or image objectives.

Because the package functions as a product's environment, it influences people's ideas about, and acceptance of, the brand. This influence can be measured in several ways. The specific techniques used depend on whether the test is intended as a preliminary or the final evaluation, and whether or not graphic design is involved.

Communications measurements can be combined with those for evaluating visual, functional and esthetic aspects of the package. The purpose of a communications test is to determine how the package changes ideas about the brand. The key to doing this is to ask questions related to the product and brand, while the respondent's attention is focused on the package—but to save direct packaging questions until the end of the interview.

Answers to the product and brand questions are influenced by what is seen. Once people are asked directly about the package they will still be able to tell you about its esthetic appeal, but the underlying influences are lost.

Some of the techniques for measuring packaging communications include use tests, advertising concept tests, and experimental studies. In new product development, sketches of packages can be used to lend reality to a verbal description, and to determine differences between what people thought the product was and what the manufacturer intended.

In a *use test*, people are given the product to try. Sometimes the trial is minimal, consisting of a single use: opening and tasting candy, opening and pouring motor oil, opening and guiding a caulking compound. Long-term use tests are often needed for products where reclosure is important such as crackers, detergents, glue and liquid medicines.

Graphic designs may be evaluated in the form of an *advertising concept test,* where the primary focus of the ad is a detailed picture of the package.

Regardless of the overall approach, realistic final studies geared to measuring the effectiveness of alternative packages need to be structured as *experimental studies.*

In this type of study, matched sets of people are asked about the test packages, with each set asked about only one packaging alternative plus competition. The characteristics of the people interviewed are similar, questioning is the same and the competitive products are the same. Since the only difference is the test package, differences in results can be attributed to the packages rather than other factors. This type of study is also called a *monadic test,* because individuals are exposed to only one of the test packages.

For edible products, remedies and the like, different types of packages and graphic designs can be treated as product formula variations. Respondents can then be asked to select the one version they would most like to try, ones they think would have certain characteristics, and ones that would be suitable for different types of people.

If possible, proposed copy and illustrations should be tested before the final package is developed since both influence product acceptance and can make a difference in communication. Alternative pictures of the contents of children's toys and games, instructions for installing equipment, generic descriptions of automotive products, recipes and the like, can change expectations for the brand, and make the difference between marginal and full acceptance.

When planning research for packaging, there are many options to choose from. Regardless of the choice, the following guidelines will get more for your packaging research dollar:

1. Determine if the packaging dimensions of interest are measurable. Subtle differences in graphic design may not be measurable, especially where an image change is intended. Some packaging objectives such as "generate trade excitement" cannot be measured.

2. Be sure the study objectives are related to the packaging goals or criteria. If a redesign is intended to increase sales and there are no other objectives, only a sales test is appropriate. The general idea is to use the packaging goals as a guide to determining what aspects will be measured.

3. Prepare a clearly written statement of research objectives, so there is

no misunderstanding about the kind of information which will be obtained or how it will be used.

4. Review the study plan to be sure results will provide the information required by the research objectives.

5. If alternative actions will be taken based on the results, set standards of acceptance before the survey starts, and be sure these standards are related to the packaging goals. This will ensure that others in the company will agree to how results will be judged.

6. Interview current or potential customers, people who can provide meaningful information.

7. Use appropriate and equivalent exhibit materials to avoid biasing results.

8. When developing packaging communications information, ask about the product and brand before asking any direct questions about the package.

Other guidelines apply to all surveys. The questionnaire should be interesting, nonrepetitive, and in spoken, rather than written English. The order of questions should not influence the answers. At the same time, there should be a logical flow to the subjects covered. Also, it is important that the wording of each question is neutral so that you do not "ask" the answer.

As in other surveys, questions have to accommodate the way the results will be analyzed. It is often desirable to use statistical techniques to increase the usefulness of a study, but this requires that questions be asked using certain formats.

To ensure that results are used, the information needs to be reported so people can understand it, and can follow the reasoning.

Designers and package developers should be involved in the research for at least three reasons:

- to be sure the research plan or proposal is compatible with the packaging objectives
- to control the appearance of test packages and pictures so these exhibit materials are appropriate for the test
- to add insights in explaining or interpreting test results

Corporations often have limited experience with design research and may not be aware of the special considerations related to these studies. For this reason, knowledgeable package developers who participate in the planning and analysis stages of research can enhance the value of the study to their clients as well as themselves.

In conclusion, the package is an integral part of most products and can provide an extra competitive edge. Successful package development is the result of careful consideration of alternatives before making informed decisions. As each packaging decision is made, it narrows the options to be considered at the next step.

Research helps those involved in package development make the informed choices that result in a successful package.

LORNA OPATOW

Lorna Opatow is president of Opatow Associates, a marketing research firm she started in 1963. Her company specializes in qualitative and quantitative research for marketing communications and design, with special focus on new product development, packaging and consumer issues. She has served on the boards of the American Marketing Association, the American Association for Public Opinion Research, Advertising Women in New York, National Home Fashions League, The Packaging Institute, and The Society of Consumer Affairs Professionals in Business.

Packaging,
Solid Waste,
and Environmental
Trade-Offs

Though explicit commitment to environmental values became an essential part of doing business by the 1980s, business responses to environmental issues often remain confused. Nowhere are the complexities of the environmental issues that give rise to confusion among business decision makers more evident than in the mundane world of packaging and waste. Solid waste and packaging issues raise the entire spectrum of tensions that make "doing the right thing" a complex affair.

Design, Production, and Marketing: Decision Dilemmas

Consumer attitudes and demographic trends, mismatches between consumer perceptions and environmental reality, and conflicts among competing environmental goals all make "doing the right thing" in product design, production, and marketing ambiguous.

Consumer Attitudes, Demographic Trends, and Buying "Green"

Despite paying lip service to environmental values, many consumers are unwilling to purchase (or purchase at a premium price) goods with notable environmental attributes (however these are defined). In this instance, pursuit of profits may conflict with commitment to putting some products with reduced environmental impacts into the marketplace.

This chapter is adapted from a longer version, "Packaging, Solid Waste, and Environmental Trade-Offs," which appeared in *Illahee: Journal for the Northwest Environment,* vol. 10(1).

This conflict is more pronounced than broad-brush surveys of consumer attitudes toward environmental issues would suggest. A 1991 Roper survey found that, on average, consumers will pay a 4.6% price premium for "environmentally sound products" (Roper Organization, 1992). Yet even this modest premium may overstate the actual willingness of consumers to pay a premium. The same survey showed that in real (rather than nominal) dollars, consumers were willing to pay a premium of just 0.1% for plastic packaging with less material than a traditional alternative; 0.2% for paper with recycled content; and virtually no premium for plastic packaging with recycled content (Roper Organization, 1992).

Other surveys likewise show ambiguous results and significant variation among different consumer groups and among different products (Green MarketAlert 1992a, 1993b).[1] A 1993 Cambridge Reports/Research International survey showed 21% of respondents willing to pay a total premium in 1993 of $41 or more per month for environmental protection (a decline from 25% in 1992). In all categories except garden products, recycled paper products, and household cleaners, over 60% indicated that they would pay either "no premium" or not more than a 5% premium on "green products" (Cambridge Reports/Research International, as reported in Green MarketAlert, 1993c).

Comparing what consumers say they will buy with what they actually buy shows a gap between words and action. This discrepancy is the so-called "halo effect," whereby "consumers talk a better story than they live. . . . The 40 to 50 percent of consumers who say they 'buy green' do so only on occasion" (Green MarketAlert, 1992a). In particular, many don't buy "green products" in the grocery category, which includes many of the packaged goods that enter the municipal waste stream. A survey by the Grocery Manufacturers Association showed that 44% of consumers reported nutritional issues as most important in their purchasing decisions (Peter D. Hart Research Associates, Inc., 1992). Some 19% reported that price was the most significant determinant of their purchasing. Only 14% cited environmental characteristics as important. Evaluating the survey, the association commented that "few grocery shoppers take the environmental consequences of packaging into account when they purchase food products; however, they strongly support recycling" (Peter D. Hart Research Associates, Inc., 1992).

Changing demographic patterns can also work against some "green marketing" efforts, such as introduction of bulk packaging or reusable containers. Four trends in particular lie behind the demands for single-serve and ready-to-eat packaged food: (1) an increasing number of single-person households, (2) an increasing percentage of two-income households, (3) a decline in the number of children per household, and (4) increasing average age of the population (Franklin Associates, 1992a).

Mismatches Between Consumer Perceptions and Environmental Reality

The second broad arena of potential tension for businesses pursuing environmental goals materializes when consumer perceptions about what is "environmentally friendly" are at odds with "reality." In this instance, the business decision maker faces a choice of succumbing to consumer misperceptions or attempting to disabuse consumers of these perceptions through educational and marketing efforts.

Misperceptions are particularly pronounced regarding solid waste and packaging. In September 1990, *Food & Beverage Marketing* reported that respondents to a survey believed that plastics made up 60% of the waste stream, though plastics actually make up only 14.3% of container and packaging waste (Figure 1) and 8% of total municipal waste. Further, respondents thought paper made up about 18% of the waste stream; in fact, it amounts to nearly 50% (Ottman, 1991). Sixty-seven percent of respondents in a May 1990 industry survey identified fast-food polystyrene (plastic) packaging as a major waste problem, though such discards make up less than 1% of municipal waste (Ottman, 1993; Franklin, 1993).

The celebrated decision by McDonald's to replace its polystyrene "clamshell" hamburger packaging with a paper wrapper illustrates the dilemma corporate decision makers face when confronted with a perception-reality gap. On the one hand, some vocal McDonald's customers perceived the polystyrene foam clamshells as "environmentally unfriendly." Yet life-cycle assessments of plastic fast-food packaging showed that they compared favorably with competing paper alternatives in terms of total energy use, atmospheric emissions, most waterborne wastes, and industrial solid waste. Regarding post-consumer solid waste,

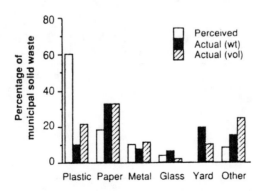

Figure 1. Perceived versus actual proportions by weight and volume of components in municipal solid waste. Source: Gerstman + Meyers Inc (1991) and Franklin Associates (1992c).

the polystyrene containers generated less waste by weight; the paperboard generated less by volume (Franklin Associates, 1990a).

The aseptic package, or rectangular juice box, offers another illustration of this mismatch between consumer perceptions and environmental reality. The state of Maine banned use of the aseptic juice box (with exceptions for a few products) as a result of perceptions by some consumer activists that the box was not recyclable and, therefore, represented wasteful packaging.

Yet a life-cycle analysis of seven orange juice packaging delivery systems shows that the aseptic requires less total energy than the six alternatives, largely because all alternatives other than the aseptic container require refrigeration: refrigeration makes up from 45% to 79% of total energy requirements in the juice delivery systems. Even at high levels of recycling for glass or plastic containers, energy requirements still exceed those of the aseptic because of the refrigeration requirement (see Table 1).

The perception-reality gap puts business decision makers into the position of either: (1) doing battle against consumer perceptions with an information campaign, or (2) acceding to their demands, even when doing so might actually move the company toward products that have greater impacts. However there is a third alternative: Find a substitute that actually reduces waste, energy use, or some other set of environmental impacts. This is, in fact, the path McDonald's ultimately took. McDonald's introduced a paper wrapper package that, although not recyclable, compared favorably in life-cycle assessments against the polystyrene clamshell and other alternatives (Environmental Defense Fund, 1991). This option did involve some trade-offs for McDonald's and for the general consumer: less heat retention with the new package, for example.[2] Moreover, even though McDonald's succeeded in finding an option that improved the en-

Table 1. Energy requirements for delivery of 1,000 gallons of orange juice (million Btu).

	0% Recycling	100% Recycling
96-oz. Plastic bottle	103	96
128-oz. Plastic bottle	102	95
64-oz. Gable carton	75	73
8.45 oz. Aseptic brick	30	27
12-oz. Composite can	116	116*
10-oz. Glass bottle	62	59

*100% composting.
Source: Coca-Cola Foods, Houston, Tx, Dec. 1989 (information used with permission).

vironmental profile of its packaging, cost-effective alternatives are not always available in the short term.

Yet the education option is also perilous. Businesses taking this tack run the risk of appearing resistant to making changes presumed by consumers to be environmentally preferred. Moreover, efforts at "educating" the public on environmental matters have had ambiguous results.

Environmental Goals: Conflicts and Opportunities

However, perhaps the most challenging environmental decisions emerge from a third source of tensions: when pursuit of one environmental goal conflicts with other environmental goals, or with efforts to satisfy other consumer values, such as safety. In this instance, the issue becomes one of deciding which values to maximize, a choice that may not always favor a particular set of environmental values.

An October 1992 report on green product design by the U.S. Office of Technology Assessment (OTA, 1992) cautions that "what is 'green' depends strongly upon context." The report expands on this perspective, suggesting that "green design refers not to a rigid set of product attributes, but rather to a decision process whose objectives depend upon the specific environmental problems to be addressed" (OTA, 1992). The report suggests that "green design" ought to be perceived as an incorporation of environmental design objectives into the broader set of design objectives that include such traditional elements as safety, product performance, product cost, and so on. Actual choices regarding product design and manufacture will depend, among other variables, on local conditions, the intended use of the product, cost, and availability of substitutes that retain product performance.

Environmental Profiles: The Problem of Trade-Offs

These choices are subtle, making even the concept of a "right choice" elusive. The OTA report offers one example: potential trade-offs between recyclability and waste prevention (OTA, 1992). To illustrate this point, the OTA report describes the modern snack chip bag, which is made up of thin laminated layers of nine lightweight materials, each of which serves a different function in ensuring overall product integrity and consumer utility. This multilayering makes recycling difficult. However, the package "is much lighter than an equivalent package made of a single [recyclable] material and provides longer shelf life, resulting in less food waste" (OTA, 1992). The aseptic package described earlier presents a similar potential trade-off between waste reduction and recyclability.

Yet these equivocal environmental results reflect only one aspect of trade-offs. For example, at fast-food and institutional food-service outlets, the choice between reusable and disposable ware presents environmental-safety trade-offs. A study of disposables and reusables in twenty-one food-service programs reported that "the probability of microbial contamination was found to be 50% greater with the reusables than with the disposable items used in the same establishment" (Felix et al., 1990). Moreover, the same study found that 15% of the reusables actually had microbial counts higher than the maximum recommended level of 100 colonies per utensil.[3]

These kinds of trade-offs among different consumer values are pervasive and, since they involve individual value rankings, cannot be resolved by appeals to science. Moreover, the nature of the trade-offs illustrates the ambiguity of the concept of "waste."

Dematerialization: A Win-Win Situation

While these tensions underlie most decisions about production and marketing, environmental and business values are not always at odds. The perennial business pursuit of reduced costs often translates into "dematerialization"—a reduction in the energy and raw material inputs needed to manufacture and deliver a given unit of output, a process that the U.S. Office of Technology calls an "environmental triumph" (OTA, 1992). This drive to reduce costs can translate into efforts to lower waste disposal expenditures by reducing "waste."

Dematerialization (source reduction, when referring to packaging) in the packaging industry has been extensive and persistent. These efforts have resulted in a reduction in packaging as a percentage of the waste stream, despite increased per capita product consumption.

Four trends account for this reduction: (1) substituting light for heavier packaging materials (for example, substituting plastic for glass); (2) introducing flexible packaging to replace rigid packaging; (3) expanding use of bulk packaging for selected products; and (4) lightweighting—reducing the amount of material used in manufacturing a given package type.

The soda can offers a compelling example of lightweighting. In 1935, the 12-ounce steel can required 235 pounds of metal per 1,000 cans, or 3.5 ounces per can. Today, the 12-ounce steel can requires 67 pounds per 1,000, or about 1 ounce per can. Introduction of aluminum cans offered additional opportunities for material reductions. Upon its first appearance in 1963, the aluminum soda can required 54.8 pounds per 1,000 cans. By 1989, however, only 35 pounds of metal were required per 1,000 units, with each 12-ounce aluminum can now weighing just over 0.5 ounces (Teasley, 1990).

Total packaging systems involve numerous parameters such as container dimensions and volume, label dimensions, transportation case dimensions, case coverings, numbers of cases per truckload, numbers of pallets per truckload, total shipping weight, and warehouse space.

All of these variables offer subtle source-reduction opportunities and can affect total energy use, materials usage, and packaging waste. Even small changes can result in significant resource conservation. In one example, a 16% reduction in the cubic dimensions of a juice package, coupled with a 10.7% reduction in label size, saved nearly 20,000 pounds of material for one producer, over 500 truckloads of outgoing freight, over 20,000 pallets, over 7,000 pounds of stretchwrap, and over 250,000 square feet of chilled warehouse space (Scarlett, 1993). These unglamorous achievements are typically lost in present-day discussions of packaging and the environment, however.

The OTA report concludes that "green design may depend on such factors as: the length of product life; product performance, safety, liability; toxicity of constituents and available substitutes; specific waste management technologies; and the local conditions under which the product is used and disposed."

Evaluating Trade-Offs

No amount of refinement in business decision-making processes can eliminate the need for making trade-offs; for example, among costs, safety, health considerations, energy and raw materials consumption, practicality, convenience, aesthetics, and so on. This complexity suggests that "doing the right thing" is best construed as development of decision-making processes that allow for explicit evaluation of such trade-offs.

Government Affairs and Environmental Policy: Ambiguous Choices

Value trade-offs associated with packaging decisions have two important implications for public policy. First, market institutions and market pricing are likely to provide useful information about resource trade-offs and relative costs of satisfying competing values; and second, flexibility for the manufacturer is critical to ensure a balancing of multiple values and the ability to make trade-off decisions where values cannot simultaneously be pursued.[4]

Command-and-Control vs. Market Responses to Price

Regulations that set specific waste-reduction and recycling standards are inappropriately patterned after legislation that sets standards for air and water emissions. Such regulatory constraints must be met regardless of trade-offs with other consumer values such as safety, convenience, and so on. The "OTA uses the phrase 'green design' to mean something qualitatively different: a design process in which environmental attributes are treated as design objectives, rather than as constraints. A key point is that green design incorporates environmental objectives with minimum loss to product performance, useful life, or functionality" (OTA, 1992).

Manufacturer Flexibility

The complexity of packaging decisions suggests that flexibility is a crucial policy consideration, allowing manufacturers to weigh the overall implications of different options. The central problem faced by decision makers in business is how to set priorities when confronted with multiple, sometimes competing, goals. As a result, "doing the right thing" involves maintaining or enhancing both internal corporate and external economic, legal, and political decision-making institutions that provide feedback about relative scarcities of resources (including those resources not fully internalized in the costs of current market transactions). At least within the limited frame of generating ever-improving efficiencies in resource use, this need for rapid feedback about relative scarcities likely means letting competitive markets and market prices function.

The importance of competitive markets, pricing feedback, and flexible responses to changing resource constraints and consumer preferences suggests that, for businesses acting in the policy and public affairs arenas, "doing the right thing" ought not to be understood as simply adopting and embracing current policy fads. Instead, in a policy context, "doing the right thing" has at least three components.

1. "Doing the right thing" includes a commitment to advancing "truth," that is, at a minimum, supporting policy based on good science. Yet, advocating good science may require an uncomfortable plunge into "data battles" and conflict, heightening consumer perceptions of corporations as being unwilling to support environmental goals.

2. "Doing the right thing" requires a commitment to weighing costs of a proposed policy against benefits. This is both a "bottom-line" and a moral issue, since dollars spent to pursue one goal cannot be spent on other goals that affect quality of life.

3. "Doing the right thing" requires exploring the institutional impacts of specific policies. Do the policies put in place institutional incentives to reduce environmental impacts? Do the institutions promote innovation and least-cost flexible responses? Do the institutions minimize transaction costs?[5]

This three-pronged set of policy guidelines is consistent with an attempt to make environmental values part of a cluster of design objectives, rather than efforts to mandate specific outcomes that necessarily incorporate a "one-size-fits-all" and "once-and-for-all" set of solutions to environmental problems.

Endnotes

1. Defining what constitutes a "green" product poses significant problems. For its analysis of sales figures, *Green MarketAlert* uses consumer perceptions as the determinant. This means green product sales can include such items as single-serve food items in containers made partly of recycled content. Whether such items, relative to alternative bulk-packaged food products, represent notable environmental benefits is subject to debate.

2. In an analysis of the McDonald's clamshell decision, economist Peter Menell notes that the most sensible decision, considering both economic and environmental factors, might have been simply to keep using the polystyrene clamshell. However, this likely would have resulted in a continued opposition to the clamshell from some segments of the public. Menell concludes, "this predicament illustrates the complex political economy of green consumerism within the current climate of hierarchical thinking, single-issue politics, and hardball advocacy" (1993).

3. Other studies have suggested that foodborne disease is a major health problem, accounting for possibly more than 12 million cases per year. Contaminated food utensils have been "found to be a significant cause of foodborne disease" (Felix et al., 1990).

4. The Office of Technology Assessments notes, for example, in Green Products by Design, "the environmental evaluation of a product or design should not be based on a single attribute, such as recyclability." The OTA report adds, "policies to encourage green design should be flexible enough to accommodate the rapid pace of technological change and the broad array of design choices and trade-offs" (OTA, 1992).

5. These three pillars of "good" policymaking pose dilemmas for business strategists. For example, they do not always produce results that are, in the near term, best for the "bottom-line." Instead, rent-seeking strategies—efforts to enhance profits by political action—may yield a higher return, at least in the short term. One manufacturer of a glass-coating material, for example, has pushed for legislation to ban colored glass from the marketplace (*Green2000*, 1991). The ostensible reason for such a ban is to make recycling easier, since clear glass is more readily marketed than glass cullet contaminated with brown or green glass. However, such bans would also create a ready market for the firm's glass coatings. The firm's strategy uses environmental rhetoric about recycling to pursue self-interested marketing goals. This strategy reinforces the general public misperception that recycling is always the preferred waste-handling option. Such a ban is not likely, however, to serve broader environmental goals because it acts as a design constraint, preventing evaluation of trade-offs among different alternatives in order to maximize overall environmental improvements.

References

Achenbach, J. 1991. On Earth Day, getting to the bottom of manufacturers' claims. *Washington Post,* April 22.

Alter, H. 1989. The origins of municipal solid waste: The relations between residues from packaging materials and food. *Waste Manage. & Res.,* 7:110.

Alter, H. 1991. The future course of solid waste management in the US. *Waste Manage. & Res.,* 9:3–20.

Angus Reid Group. 1991. Environment U.S.A. Toronto, Canada.

Arthur D. Little, Inc. 1990. *Disposable versus Reusable Diapers: Health, Environmental and Economic Comparisons.* Cambridge, MA.

Ausubel, Jesse H. 1989. Regularities in technological development: An environmental view. In Jesse Ausubel and Hedy E. Sladovich, eds. *Tech. and the Environ.,* National Academy Press, Washington, DC, pp. 70–91.

Environmental Defense Fund—McDonald's Waste Reduction Task Force. 1991. Final Report. Environmental Defense Fund, Washington, DC and McDonald's Corp. Chicago.

Felix, Charles W., Chet Parrow and Tanya Parrow. 1990. Utensil sanitation: a microbiological study of disposables and reusables. *J. of Env. Health,* 2:13–15.

Felix, Charles. 1990. Foodservice disposables and public health. *Dairy, Food and Env. Sanitation.* Nov. 1990: 656–660.

Frankel, Carl. 1992. Blueprint for green marketing. *American Demographics Magazine.* April 1992. Ithaca, New York.

Franklin Associates. 1989a. The role of beverage containers in recycling and solid waste management. Prairie Village, Kansas, report prepared for Anheuser-Busch Companies, St. Louis, MO.

Franklin Associates. 1989b. Energy and environmental profile analysis for orange juice container systems. Report prepared for Coca-Cola Foods. Houston, TX. Dec. 1989.

Franklin Associates. 1990a. Resource and environmental profile analysis of foam polystyrene and bleached paperboard containers. Prairie Village, KS. June 1990.

Franklin Associates. 1990b. Disposable diapers: summary and interpretation of literature sources on the environmental and health effects of diapers. Prairie Village, KS. July 1990.

Franklin Associates. 1992a. Analysis of trends in municipal solid waste generation: 1972–1987. Prairie Village, KS. January 1992.

Franklin Associates. 1992b. Energy and environmental profile analysis of children's single-use and cloth diapers. Revised report. Prairie Village, KS. May 1992.

Franklin Associates. 1992c. Characterization of municipal solid waste in the United States: 1992 update. July. EPA-530-R-92-019. PB 92-207 166, National Technical Information Service, Springfield, VA.

Franklin, Marge. 1993. Franklin Associates. 4121 West 83rd St., Suite 108, Prairie Village, KS, interview, 2 August.

Green2000. 1991. Newsletter published by Packaging Strategies, Inc., West Chester, PA. August 1991.

Green MarketAlert. 1992a. May. Newsletter published by The Bridge Group, Bethlehem, CT, May 1992.

Green MarketAlert. 1992b. November.

Green MarketAlert. 1993a. February.

Green MarketAlert. 1993b. April.

Green MarketAlert. 1993c. October.

Herman, R., S. Ardekani, and J. Ausubel. 1989. Dematerialization. In J. Ausubel and

E. Sladovich, eds. *Technology and Environment*. National Academy Press, Washington, DC, pp. 50–69.

Lehrburger, C., J. Mullen, and C. V. Jones. 1991. Diapers in the waste stream: A review of waste management and public issues. Report for the National Association of Diaper Services. Great Barrington, MD.

Makower, Joel. 1993. *The e Factor: The Bottom-Line Approach to Environmentally Responsible Business*. Times Books. New York.

Menell, P. 1993. Eco-information policy: A comparative institutional perspective. Unpublished manuscript. University of California, Berkeley School of Law.

Office of Technology Assessment, U.S. Congress. 1992. *Green Products by Design: Choices for a Cleaner Environment*, OTA-E-541. U.S. Government Printing Office. Washington, DC, October 1992.

Ottman, Jacquelyn. 1991. *Environmental Consumerism: What Every Marketer Needs to Know*. J. Ottman Consulting. New York.

Ottman, Jacquelyn. 1993. *Green Marketing: Challenges and Opportunities for the New Marketing Age*. NTC Business Books, Lincolnwood, IL.

Peter D. Hart Research Associates, Inc. 1992. Opinion 92: America's Consumers Speak Out. Survey prepared for Grocery Manufacturers of America, Washington, DC.

Rathje, William, and Michael Reilly. 1985. Household garbage and the role of packaging. Unpublished manuscript. University of Arizona. Tucson, AZ.

Rathje, William. 1989. Rubbish! *The Atlantic Monthly*. December, pp. 99–109.

Roper Organization. 1992. Environmental behavior, North America: Canada, Mexico, United States. Survey commissioned by S. C. Johnson & Son, Inc.

Scarlett, Lynn. 1991. The garbage crisis: Air, water, earth, and people. Paper presented at the Mont Pelerin Society Regional Meeting, Bozeman, MT, August 1991.

Scarlett, Lynn. 1993. Mandates or incentives? Comparing packaging regulations with user fees for trash collection. Reason Foundation. Los Angeles, CA.

Teasley, Harry. 1990. Thinking about recycling. Paper presented to the Recycling Advisory Council, Washington, DC, October 23, 1990.

Teasley, Harry. 1993. Keeping free enterprise free. Presentation to the University of Tampa Fellows Forum. Tampa, FL, February 19, 1993.

LYNN SCARLETT

Lynn Scarlett is Vice President of Research of the Reason Foundation, a nonprofit, public policy think tank based in Los Angeles, California. Scarlett has a B.A. and M.A. in Political Science from the University of California, Santa Barbara. She completed her Ph.D. coursework and exams (with dissertation in progress) in Political Economy at the University of California, Santa Barbara.

Scarlett has written extensively on environmental policy issues, with a particular emphasis on natural resources, solid waste, recycling, and air quality. Her work has appeared in such journals as *Regulation* magazine, *Journal of the Northwest Environment, Journal of Social and Economic Regulation,* and Houghton

Mifflin's *1994 Encyclopedia of the Environment*. She is also the author of
a widely circulated booklet, *A Consumers' Guide to Environmental Myths
and Realities*, which examines popular perceptions about solid waste,
recycling, and packaging issues.

Scarlett has also written widely in the popular media, having published
articles in *The Wall Street Journal*, the *Los Angeles Times*, and *Reader's
Digest*, among other general-audience publications. Scarlett has appeared
three times on ABC's Good Morning, America and twice on CNN's Cross-
fire to discuss environmental policy and recycling.

Scarlett is currently chairing a "How Clean Is Clean" Working Group,
sponsored by the Washington, DC-based National Environmental Policy
Institute. She was appointed in 1994 by Governor Pete Wilson to chair
California's Inspection and Maintenance Review Committee, charged with
making recommendations to the state legislature regarding California's
Smog Check program. Scarlett serves as technical advisor to the Solid
Waste Association of North America's Integrated Waste Management Project; and she served as technical advisor to Houghton Mifflin on its Encyclopedia of the Environment project.

The Cons and Cons
of Copycat Packaging

A small but significant change in marketing strategy was obvious at the 1994 Private Label Manufacturers Association (PLMA) trade show in Chicago. Compared to 1993 when store-brand package designs were displayed alongside national-brand packages in order to point out the remarkable design similarities, at the 1994 show the national-brand products were nowhere to be seen.

I saw this as a clear response to the legal action and criticism that private-label manufacturers had recently experienced for what was seen as their misappropriation of national-brand package designs. Exhibitors at the PLMA show suddenly seemed uncomfortable touting their copycat containers in direct comparison with those used by national brands. Yet copying rather than innovating as a tactic for store-brand products seemed not to have diminished at all.

For instance, only within a few weeks preceding the show Warner-Lambert's new squarish bottle for Listerine became available nationally. But in two different booths at the PLMA show, I saw near-duplicates of that distinctive container, ready for filling with a store-brand mouthwash. I've been told that Warner-Lambert's new design incorporates some difficult-to-copy details, but the knock-off bottles looked close enough to create consumer confusion.

No doubt distressing for McNeil Consumer Products, maker of Tylenol analgesic, near-duplicates of the FastCap easy-off closure also appeared at the PLMA show. As I understand it, McNeil made a serious error in not patenting that design.

There were examples of store-brand containers that had been redesigned so as to differentiate themselves from the national brands. A bottle

35

for a store-brand version of P&G's Pantene shampoo, for example, had gone to white from a P&G-like pearlized ivory, and its former swirl-like symbol had been changed to a "scribble," which bore little resemblance to P&G's swirl.

Nevertheless, a visit to a supermarket where store-brand packaged goods are sold remains a confusing experience. National-brand products are displayed shoulder-to-shoulder with private-label copycats. In size and shape, in container material, in color and labeling, each national-brand SKU and its P-L clone all too often look enough alike to cause a Hollywood-perfect double-take. But this caper is not funny. A store brand's copycat packaging can harm a national brand's hard-earned brand loyalty. And that's exactly why national-brand companies increasingly are taking store-brand manufacturers and retailers to court.

Attorneys hired by the national brands are trying to and succeeding in finding consumers who have purchased private-label products in error. Although these cases of mistaken identity are costing national-brand companies millions of dollars in legal fees, defense of a proprietary package design is crucial if brand equity is to be maintained. Litigation seems certain to increase unless even the most enthusiastic supporters of private-label products start to campaign vigorously against copycat packaging.

What's the big deal? Isn't this just "good old American-style competition"?

Well, as marketers are coming to understand sufficiently, packaging is an extremely powerful, maybe the most powerful, consumer-goods sales tool. Think about it: The typical American supermarket displays 20,000 to 30,000 different items and the typical shopper spends 20 minutes zooming past them. In this era of media overload, product proliferation and market segmentation, it's no wonder consumers make eight out of every ten purchase decisions in the store, not before. Although shoppers seem reluctant to admit that packaging greatly influences their habits, it irrefutably does. A package can become so closely associated with a brand's identity that it and the product become one and the same in the consumer's mind. A unique package design can establish and strengthen a brand's equity; a private-label knock-off can quickly dilute it, destroy it.

No wonder some national-brand companies that invest heavily in getting each new package sketched, prototyped, tested and distributed are angry. Private-label marketers have been duplicating those proprietary containers, filling them with products of questionable quality, and selling them for less.

In the course of research for my monthly newsletter *Shelf Presence*, I've spoken with a number of package designers who admit to developing copycat packages for their private-label customers. Others say they have repeatedly refused, sacrificing hefty fees in order to avoid being implicated in a package-infringement case. Attorneys generally advise designers to stay away from P-L clients who request package duplication, simply because the situation is getting so nasty: As one intellectual-property lawyer told me recently, "The gloves are off."

All right, not *all* gloves. Some national-brand concerns are staying out of the ring, realizing that a jab at store-brand clones would bloody themselves: Many, perhaps most, national-brand manufacturers have private-label subsidiaries or divisions. (Now, if those national-brand companies are both supplying comparable product for private-label sale and allowing retailers to copy their packaging, what we have here is a corporate personality crisis bordering on clinical schizophrenia.)

National-brand companies that make products for private-label marketers face another damned-if-you-do/damned-if-you-don't situation: Suing a store that sells P-L items becomes a matter of biting the hand that feeds you. That's why recent legal action taken by major national-brand companies against retailers has generated such publicity.

Let's be fair. Some private-label concerns are innovators. A&P's Master Choice, Loblaw's President's Choice and Wal-Mart's Sam's Choice are among them. Such "premium" private-label lines are becoming brands in and of themselves. And they have acquired their premium status in great part through superb, original-looking packaging.

It only makes sense. If a marketer of private-label products can deliver to consumers something that is different in formulation, there is no reason to put it in a look-alike package. Put a less-expensive or more-effective shampoo in a bottle that duplicates the look of the national brand's and what have you got? Sameness. Now, take that shampoo and put it in a bottle that is easier or more fun to use than the national brand's and what happens? You acquire a reputation for superiority in every way. Sometimes all it takes is one little, relatively inexpensive, tweak.

What kind of tweaking does it take to create a really ergonomic, convenient package? How about the bottle that Pillsbury Hungry Jack Microwave Ready pancake syrup comes in? Introduced last year, this chubby jug is shaped to fit into the microwave oven for heating. It's got a handle that stays cool to the touch, a special closure that automatically releases steam and a label that says "Hot" when the syrup is ready to serve. There's nothing like it in the P-L world, nothing.

Marketers of private-label pancake syrup could have done it first. Sure, it would have been gutsy. Sure, it would have required a large investment in design and tooling. Sure, it might have proved unpopular. But business is risk. I'm convinced that taking chances with a package designer is better than taking chances with a judge and jury.

Nevertheless, many marketers of store brands appear to be holding out hoping, it would seem, to strike a deal when ordered by the courts to cease and desist. In response, national-brand marketers are becoming more aggressive. They realize they can't stop a competitor from bringing out a P-L product that is similar to their brand. The one thing the national-brand companies can do is work to prevent having the consumer buy the store brand by accident.

Although so many design-infringement cases are settled out of court that it's impossible to track wins and losses, a strong likelihood of mistaken identity seems to give the edge to the national brands.

Here's an example: In an effort to regain sales that had been lost to private-label products, a national-brand company switched from a stock bottle to a custom-designed bottle for its reformulated body lotion. A store-brand competitor, having taken notice, rushed to get a nearly identical container manufactured. Then the store-brand competitor told a graphic designer to closely approximate the national-brand label. Almost as quickly as the store brand hit the shelves, the national-brand company filed suit. The attorney hired by the national-brand company worked with the company's consumer-relations department to locate a woman who had written a letter of complaint. As it turned out, the product she had purchased and disliked was not the national brand but its store-brand clone. That was enough to convince the judge in the case that additional confusion was likely.

A number of consumer-research companies offer services that can strengthen a claim of mistaken identity. One employs people to observe shoppers as they make selections in a store, then interview them about what they think they purchased. Sometimes, the consumers realize only then that the product they believed to be the national brand was in fact a private-label approximation. Other researchers provide highly sophisticated systems that can, for instance, track a subject's eye movement as he or she views an array of packaged goods on a store shelf. Results of such studies can be helpful in proving that one package design is confusingly similar to another.

But all of this happens after the rip-off artist has done the damage. There's got to be a better way, and there is. Companies planning a package-design project must do their homework. They must look to the past as much as to the future, developing a sense of history along with a vision of things to come. And they must put a lawyer on their new package-development team at the start. (Consider it a cheap insurance policy.) Then it's a matter of choosing a package designer who is creative, ethical and appreciative of solid market research.

Here's a tip: Use some element on your package that would be extremely difficult for anyone to duplicate without great legal risk. Sophisticated printing, a unique logo or typeface and an unusual container shape are just three examples. The more difficult it is for a store-brand copycat to approximate such a feature, the more determined the store brand may be in attempting to do so. But the closer they get, the bigger their gamble.

A scholarly study of national-brand/store-brand competition published in the October 1994 edition of the *Journal of Marketing* (American Marketing Association, Chicago, IL) comes to a conclusion that I couldn't agree with more. Authors Paul S. Richardson, Alan S. Dick and Arun K. JainÑall university professors of marketing, finish their complex article with the following statement: "Active marketing of store brands implies investments in creating a high-quality image for these products and a commitment to offering a level of real quality that is equal to or surpasses that of national brands. Such a strategy entails use of imaginative and aesthetically pleasing package designs that differentiate store brands from the competition and prompt impulse purchase."

With regard to national brands, the professors say this: "Marketers of national brands may note that success depends on not only maintaining a high level of intrinsic product quality but also making investments to develop a strong brand image. When there is no difference from a performance point of view, consumers may be less willing to pay the premium for national brands."

The stakes are high. A retailer who develops a strong line of store-brand products can do one thing a national-brand company cannot: acquire a competition-killing presence in every aisle of the store. Taking people to court is expensive. But so is developing a unique package and sitting back doing nothing while it is copied by unscrupulous competitors.

Even those store brands that replicate national-brand packaging down to the millimeter can, if the product inside is good and the price is low, offer the consumer a desirable alternative. Private-label doesn't have to keep its knock-off image. It can change its attitude and its packaging strategy. And it will have to, with or without a judge's orders, simply because the national brands are going to keep investing in periodic package updates.

No matter who changes first, there's an upside to all this as far as the consumer is concerned. Whenever a company decides to revamp its packaging in a game of one-upmanship with a competitor, an opportunity to excel presents itself. Skillful research can lead to packaging that not only wins design awards but solves consumer problems. And that's the best marketing strategy I know of.

GREG ERICKSON

Greg Erickson is editor/publisher of *Shelf Presence*, Buffalo Grove, Illinois, a monthly newsletter-format resource guide to "creating the package that sells the product." Its circulation includes leading consumer-goods companies, market-research professionals and package designers. Erickson formerly was editor in chief at *Packaging* magazine. He is a member of the Institute of Packaging Professionals and the American Marketing Association.

Private-Label Packaging:
From Wallflower to
"Belle of the Ball"

In the 1990s, private-label packaging has become a strategic tool for retailers and wholesalers in the U.S., allowing them to differentiate their business from that of the competition. Working alone, with packaging designers, and/or with manufacturers, retailers and wholesalers have developed truly distinctive, sometimes award-winning private-label packaging, including innovations such as flexible paper packaging for baby diapers, which was introduced early in 1994 by Arquest, Inc., Cranbury, NJ, a dedicated private-label manufacturer.

Prior to the 1990s, private-label packaging in the grocery trade in the 20th century had evolved from being merely a packer-label substitute into full brand status. Traditionally, a private-label product has been easily identifiable by its packaging: It was dull, lifeless and mostly functional; that is, serving to protect and identify the product inside. Virtually no aesthetic effort was made to enhance its appearance or inform the consumer about any product features. Frequently, private-label packaging was an afterthought, drawing on used stock materials to keep costs down.

From the 1950s into the 1970s, labels in private-label programs proliferated. Most of the private-label stock was positioned at a lower price point than nationally advertised brands, so its packaging expressed this cheaper or bargain-brand image. The packaging was uninspired, often provided by suppliers, using stock vignettes, and printed in a rubber-stamp fashion. This resulted in two basic strategies: the family look or a common motif across all product categories or separate, often unconnected identities assigned to different product categories—a hodgepodge.

In the early 1980s, as private-label programs expanded with more health and beauty care products being added, efforts developed to copy the leading brand look in each product category in order to suggest that a quality comparison could be made. It was, in effect, the first industrywide move toward "artistic treatment" of private-label packaging, paying more attention to container shape, brand-name identity, graphics, colors, symbols, type styles, mass display effects, secondary packaging, closures, and general package functions.

In health and beauty care the trick was to copy all of the national-brand strategies in packaging, but not to trick or confuse the consumers into thinking the private label was the national brand. Frequently, comparisons were made—sometimes right on the private label packaging itself. The strategy was to relate to the national brand as closely as possible, but still maintain a private-label identity. This trend led to private label being stereotyped more as a copycat brand.

Private-label packaging also gained more attention from retailers and wholesalers with the debut of generics in the late 1970s and early 1980s. These products, appearing mostly in plain black-and-white or yellow-and-black packaging, with little or no graphics, addressed the lower-quality tier. In a sense, generics gave first-tier private label an opportunity to upgrade its look and distinguish itself as being of a better quality than the generics—in fact, equal to or better than the leading brands. Nevertheless, all private labels again were stereotyped, this time as "the generics" or no-name brand. Even though private label did carry a store-brand identity in improved packaging, this derogatory tag stuck because private label really did not carry the weight of a brand name per se: A brand name connotes recognition by the consumer who is exposed to national-brand advertising and promotions.

However, through the 1980s private label began to attract more consumer recognition, respect and loyalty, helped along by its improved packaging and higher quality-contents. Package designers were retained by retailers and wholesalers to upgrade the look of their private-label lines across all product categories, giving them more of a branded presence.

Since then, packaging refinements have evolved as a result of growing awareness of the impact of packaging on sales, a general copycat mentality in the industry, and influences from retailers abroad. The industry really defined itself late in the 1970s with the establishment of a trade magazine followed by a trade association, both dedicated to private-label development and the promotion of this concept. The Private Label Manufacturers Association in New York has conducted trade conferences with educational seminars, which have helped industry people focus on the nuances of packaging. Similarly, *Private Label Magazine*, published in Ft. Lee, NJ, has worked toward educating the trade about packaging through its articles about successful private-label programs as well as articles from industry experts or reports on trade meeting speeches about packaging.

The private-label industry has not been able to shake its copycat strategies. But the orientation has changed. Name-brand leaders no longer are the only standard bearers for packaging change. Successful private-label programs now attract imitation from both competing and noncompeting private label programs. Other successes in this business are quickly imitated.

Some aggressive, forward-thinking players have worked through inhouse brokers to achieve outstanding private-label programs, including superior packaging strategies. One case in point is Ralphs Grocery, Compton, CA, who worked with the broker Daymon Associates, Stamford, CT, to develop superior Ralphs brand packaging in the early 1980s. Outstanding, four-color graphic photographs of food set against a black background gave its private-label packaging pizzazz. A number of retailers and wholesalers across the country licensed this design, putting their own logo on the packaging.

Who was behind this design concept? Answer: The Don Watt Group, Toronto, who had been working with Loblaw's in Canada on creatively introducing a private-label first-tier program under the "no name" identity. Don Watt has since become a household reference in the private-label industry, having helped Dave Nichol, then at Loblaw's, to dress his new President's Choice upscale program in truly unique package designs, starting in 1985. Both Watt and Nichol had been strongly influenced by European retailers like Marks & Spencer in the United Kingdom, Albert Heijn in the Netherlands, Migros in Switzerland, Carrefour in France, and so forth.

Other package designers in the U.S. also have helped enhance the brand status of private label through package redesign. Through the 1980s up to the present, companies like BrandEquity, Gerstman + Meyers Inc, MVP, Landor, and others have helped dramatically change the image of private label.

Private-label package design has dramatically matured within three basic quality tiers. For example, retailers and wholesalers in the industry have changed their generic packaging strategy from a mostly plain black-and-white appearance to more of a private-label look that was commonplace in the 1950s and 1960s—a dull, almost nondescript rubber-stamp logo/design treatment, suggesting a lower price image. In turn, the largest private-label programs—the first-tier quality lines that match the leading brand in each product category—have been upgraded with stylized, sophisticated, four-color packaging treatments. Major packaging design firms have helped retailers and wholesalers differentiate themselves from their competitors with distinctive designs, such as pinstripes or a stylized photographic execution.

For example, working with Landor, San Francisco, H. E. Butt Grocery Company, based in Texas, has opted for illustrations instead of product photography on its packaging. This retailer doesn't just turn the packaging assignment 100% over to Landor. Instead, H. E. Butt issues category or design briefs to Landor for each product, describing what it competes against and what should be said on the packaging in terms of a unique or spirited impression.

However, many private label programs still look to studio-quality photography on the packaging to help convey top quality. This treatment takes on different stylized approaches. Retailers like Kroger, Cincinnati, use effective, snappy graphics to liven up product appearance, including, for example, wraparound graphics on the packaging of its taco sauce and taco dinners. Many times, designers look for a truly singular mood that can be carried across different product categories. If that's not possible, the designers develop a subbrand strategy, very often tied to the flagship store brand. They look to develop packaging personalities as well as a product statement, not only on this quality level but below and above it as well.

The most creative treatment of packaging occurs in the upscale or premium private-label lines. President's Choice offers some truly unique products that carry outstanding, award-winning graphics. The same holds true for the upscale Master Choice program at A&P, which seeks to convey an image of excellence in its product range. In Canada, A&P recently captured a Gold Clio Award for its packaging of Master Choice granola bars and a Silver Clio Award for its Chicago Deluxe Pizza also under the Master Choice brand.

The packaging strategies for many upscale programs are centered around truly distinctive, eye-catching designs that can surpass the brand-name leaders. The idea is to present superior quality, more product quality, and/or unique products. This is reflected in the packaging through design and very often in the packaging "pitch," such as in Master Choice's "Extra Exceptional 40% peanuts and peanut butter chips" cookies. Sometimes this strategy takes an interesting turn, such as with Weis Markets in Pennsylvania, who also uses the brand name of one of its key ingredients in the product on the Weis Choice cookies package (i.e., oatmeal cookies made with SunMaid raisins or peanut butter and chocolate-chunk cookies made with Peter Pan peanut butter).

In some cases, the package designers use the packaging itself to convey the quality message, often with long discussions about the product. This has been a strategy with Don Watt, working with President's Choice and, more recently, Wal-Mart's Sam's American Choice products. It's also not uncommon to see recipes appear on the back panels of such packaging.

Retailers and wholesalers are now working toward more product innovation in their private-label programs, introducing new lines of product with distinctive packaging, while also addressing the latest consumer interests—healthier foods, less fat, no sugar, ethnic foods, etc. Together with outstanding packaging, the product itself helps them differentiate themselves in the marketplace. Topco, a buying co-op based in the Midwest, for example, has a Top Care European Formula line of health and beauty care products, carrying its own distinctive packaging.

Just on packaging changes alone, private label over the past fifteen years has evolved into truly an exclusive brand status.

PHILIP FITZELL

Philip Fitzell is publisher of the *Exclusive Brands Sourcebook*, published by Exclusive Brands Publications, New York. Mr. Fitzell also has recently organized an information and services Internet/World Wide Web home page, the PL-EB Interchange (http://www.incadinc.com/pl-eb) for the private-label industry worldwide. He has some thirty years' writing experience, including over fifteen years in the private-label field, having helped start *Private Label Magazine, Private Label International Magazine*, and the Private Label Manufacturers Association. Mr. Fitzell also is a frequent speaker at trade meetings around the world.

VI

Two Compelling Reasons
Why Your Package
Should be Red
and White

The package is one of the famous Ps of marketing, going along with product, price, positioning and promotion. Compared to all the attention usually given its cousins, packaging has received much less focus and discussion.

Looking at the future direction of packaging and packaging research gives me an opportunity to share my favorite story about direction and the importance of gaining a good perspective before moving in any direction. Seems a sea captain saw what appeared to be another ship approaching at *night*. He had his signal man flash "Please change your course twenty degrees south." The reply came back, "No, please change *your* direction twenty degrees south." The captain replied, "I said change your direction twenty degrees south, I'm a captain"—to which the reply came: "I'm a seaman first class, change *your* direction twenty degrees south." The captain was furious. "Dammit, I said change *your* course twenty degrees south. I'm on a battleship." Back came the reply—"and I said change your course, I'm in a lighthouse."

As we ponder the implications of our captain's "direction," let me get to the heart of my message. By some wonderful twist of fate it has been my remarkable good fortune to have spent almost my whole career to date with the two greatest brand trademarks in the world—Coca-Cola and Campbell Soup—and guess what . . . they both chose the colors red and white (see page 49).

Perhaps even more remarkable, after I chose the title of this chapter, with tongue in cheek, I remembered a number of other world-class brands that also have chosen red and white as their battle colors. What about Heinz ketchup? They get their "red" from the product itself.

Or the King of Beers—Budweiser—the St. Louis institution that has almost always been red and white.

What about the world's #1 cigarette brand Marlboro—which is packaged in a smashingly successful red and white "box" (see page 50)?

There are others—Colgate and Tylenol, for instance. Carnation is one that I'll come back to because there's an interesting lawsuit I want to comment on.

But back to Coke and Campbell. The Coke contour bottle was born in 1915, quite a chubby baby until it slimmed down in 1923. The story I'm told is that Coke wanted a shape that could be identified in the dark—talk about a proprietary package! More on the contour bottle in a minute. And what about that "New Coke" can in 1985? Well that's a separate story.

Back to Campbell Soup, the brand that lingers in my bloodstream years after my departure. Good ol' Campbell Soup. Second only to Coke in consumer-brand recognition.

Campbell's switch to red and white happened at the turn of the century. The story is that Dr. J. T. Dorrance attended a Cornell football game and fell in love with Cornell's uniform colors, red and white. I guess if he'd gone to my alma mater Penn, Campbell's would be red and blue!

At this point many of you must be asking where the power of red and white comes from. Why is it the look of the leader? There used to be a packaging guru by the name of Louis Cheskin from Chicago who was a real expert on color. Louie is no longer with us, so I turned to the Wagner Institute of Color Research. Carlton Wagner says this: "Red is the color that causes the body to pump adrenaline, making the heart beat faster. People eat and drink more and longer in the presence of red. Infants will look at red in preference to any other color. Women prefer blue-based red, men favor yellow-based."

Wow! Who would ever dream red has all these attributes. And we all know white stands for purity and quality. Put 'em together . . .

Back in the 1950s, Campbell's president sponsored some research to show how a red and white look-alike label could confuse consumers. He had the research designed to support a lawsuit Campbell brought against Carnation, which was using a red and white Campbell-like label for condensed milk! The research asked consumers to identify brands based on pictures. Some of the pictures of Campbell's soup used the name "Gong-dotte" in place of Campbell's. Seeing the red and white in the Campbell pattern, consumers identified it as Campbell's, even when the name was actually Gongdotte. The research was successful, but we lost the suit anyway (see page 51).

Campbell Soup and Coca Cola: Two reasons why packages should be red and white.

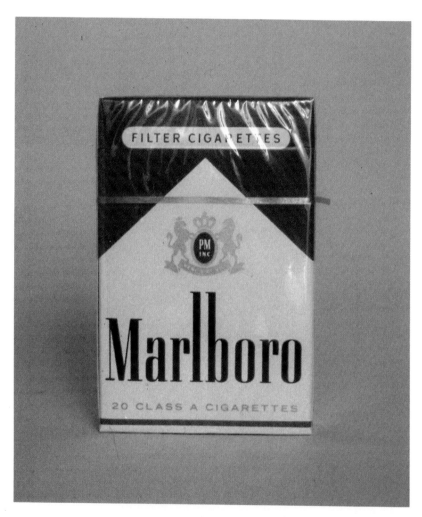

The Marlboro Box: If you need another reason.

Gongdotte was immediately identified as Campbell.

54

Research confirmation of shape equity leads to push of Coke's contour package. Even the cans include a picture of one contour bottle.

Campbell Soup finally added vignettes in spite of its icon status.

Coke recently had a similar problem—but in its own category. Look at the two Sainsbury cola bottles—looks like good old Coca-Cola, right? Not so. It's Sainsbury's store brand of cola—using an RC cola formula and a Coke bottle look-alike label (see page 52).

It took Coke six months to get this U.K. retail chain to agree to change what appears to be an obvious attempt to confuse shoppers.

A strong brand is a priceless asset, so it is not surprising that competitors try to usurp some of its value. There has been a lot of discussion about brand equity lately, and when we talk brand equity, the package has to be a primary issue—it is part and parcel of the "priceless asset." The package is the "salesperson on the shelf." The number of packaging impressions dwarfs advertising impressions. We're talking billions of impressions each year vs. millions for ad copy. And we all know that the package has to close every sale—it's that last few inches of the 1,000 mile pipeline from the manufacturer to the consumer.

When you've got great equity in your package along with a brand like Coke and Campbell, people like Andy Warhol give you millions of dollars of free publicity like Andy did for Campbell.

When you've got great equity in your package along with your brand, like Absolut Vodka, you can use your package to become your advertising message.

Maxwell House, Campbell's, Wheaties and Hershey's are all great brands that consumers can identify by color and shape alone (see page 53).

Speaking of shape, it's created package equity, not just for Coke, but for L'Eggs and Michelob. And, by the way, I don't think Michelob takes advantage of their great shape equity.

I'm sure many of you have read or seen the new push behind the classic Coke contour package. Like Michelob, Coke was taking its *shape* equity for granted. Only a fraction of total Coke consumption was in the 8 oz. contour bottle. All 12 cans and 2-liter plastic bottles look the same. They're generic to all soft drinks. But then a funny thing happened. In a new series of TV ads, the contour bottle was very prominent. Young people—teens—called and wrote, "Where can I get that great looking Coke bottle?" It was new to them. The rest is exciting history. Research confirmed a 4-to-1 preference for the contour package. Today Coke offers 20 oz. plastic contour containers, 1- and 2-liter contour packages are on the way—and they even put a picture of the contour bottle on their aluminum cans! This has to be a first. For older consumers, there has been a twinge of nostalgia. For younger folks, it's different and exciting. Sales went through the roof (see pages 54–55).

Meanwhile, back in Camden, NJ, a man named David Ogilvy was starting a mini packaging revolution of his own. One of the agency's creative and research giants of the 20th century, Ogilvy admonished Campbell management by observing that when he went to his cupboard/pantry, all the food packages said "eat me" with delicious appetite vignettes except . . . Campbell's Soup . . .

"It looks like an oil can," said Ogilvy. Well, even though the impact was earth shaking, management needed more evidence that Ogilvy's intuition was right. It took a number of years and some pinches in the rear end by yours truly to get Campbell to go to the next step . . . consumer testing. Eliot Young and his gang at Perception Research twice had to demonstrate that a change from Andy Warhol's favorite label wouldn't turn away loyal users. They did a fine job, and the result is a label with product vignettes. The amount of white space has grown significantly to accommodate the soup vignette. The research votes are in and so are supermarket dollars. It was a winning move (see page 56)!

Stimulating that kind of sales performance is the final objective of any packaging change. Before getting at that final objective, however, we need to define what I see as the two major objectives of packaging performance.

Key Package Performance Measures

1. Physical impact
   • Clutter—Does it stand out in its competitive array?
   • Shelf-impact findability—Can you find it?
2. Positioning/communication
   • Consistency with brand strategy or change of strategy
   • What benefits/attributes are communicated?
   • Is the communication *persuasive*?
   • Does the *communication* address *user imagery*?

If you buy my performance objectives, then the research criteria are pretty straightforward, measuring *impact, communication* and *persuasion*. These criteria have been around for years. What's changed is technology. We now have virtual-reality testing simulations, digitized displays, etc. In short, more realism in the research tasks.

Let me leave you with a couple of personal observations and a couple of non-Campbell/Coke examples.

First, I hope I'm not stomping on too many readers' toes, but the new brand manager routine of automatically making a package change is hurting some well-earned brand equities. If there's a legitimate need to address design deficits, great—otherwise leave the package equity alone.

My colleagues back at Campbell changed Swanson's frozen line to a blue background, yet an awful lot of packaging folks will point out what a lousy food color blue is.

Second, beware the "bean counters." The cost accountants and CFOs are looking for every dime they can save and put on the bottom-line. Unfortunately, when they go overboard on packaging, brand equity can slide out the door for the sake of one good quarter or year. Remember when all breweries were following like sheep to the cheapest glass container—the glass can? Everyone looked alike. Then someone, Bud or Miller, found out that "old-fashioned" *long* neck bottles were much preferred to drink from and, perhaps equally important, to be seen drinking from in bars and restaurants.

Pepsi thought they had a great new concept in Crystal Pepsi, but they packaged it in standard containers. Revolutionary new product—dull old containers. Meanwhile Coors went the extra packaging mile for Zima—revolutionary new beverage—totally new package look.

And someone at Chesebrough deserves a lion's badge of courage for Mentadent. The accountants and environmentalists must have been on his or her back, but the package makes a very bold statement about the quality and uniqueness of this great new product.

Magnificent or slipping, brands must stay in touch with what their packaging is contributing to their consumers' purchase decisions and be open to packaging innovations that can enhance their packaging equity and their top- and bottom-lines.

Brands must have a real and meaningful packaging strategy to match up with their product or category marketing strategy. If they don't—if they're still executing packaging with a 1970s or 1980s mind-set—they are missing out on opportunities for substantial profitable growth.

Brands must know how their packages factor into their consumers' buying decision, and they must know it in each category in which they compete, because the differences can be dramatic.

Brands should know how "elastic" consumer pricing is for packaging in the categories they compete in. Will the consumer pay more for relevant difference or is cloning the category leader the best road to profitable growth?

Brands should know what the profit and growth "drivers" are—from a consumer perspective—in their categories. Understanding the packaging implications of these drivers from a consumer perspective can often mean the difference between great overall performance and an "also ran."

Brand marketers should be aware of the reliable and valid research tools that are available to help make key package decisions. All too often, consumer research becomes a couple of focus groups in a church basement with eight or nine relatives of employees. There's nothing wrong with focus groups if they are positioned and executed professionally. But there are other tools available to you that often don't cost any more than focus groups. Tools that maximize your returns on this key "P" of marketing.

Finally, brand marketers should factor the consumer into their packaging decisions. Too often they let their financial folks and a cost-of-goods criterion dictate all of their packaging decisions. As I read the consumers of this decade, I continue to see evidence that they want to know what a product is worth, not just what it costs. If you know and understand the difference as it relates to packaging, packaging can be one of your key growth drivers for years to come. And if you are feeling hungry or thirsty after reading this chapter, it may be that the power of red and white extends to the printed page as well.

TONY ADAMS

Tony Adams is a former Vice President of Marketing Research and Planning at Campbell Soup Company, and a Director of Strategic Marketing Research at Coca-Cola. In addition to teaching at the University of Pennsylvania's Wharton School, he is a marketing consultant.

VII

No More Tilting at Windmills: The Consumer Defines Today's Packaging Systems

Over a decade ago, Mona Doyle, as President of Consumer Network, invited me and other representatives of the packaging industry to sit in on a wide-ranging consumer focus group session. The consumers on the panel discussed what they liked and disliked in packaged products and why; what they would like to see changed and how; and whether they would pay for those changes in the products they selected for purchase. At the session's end, the consumers left feeling the packaging industry was interested in their issues because we showed up—and we listened.

I later accused Mona and her consumer panelists of being somewhat like Don Quixote and tilting at windmills. Through Mona's focus group, we saw that the irritations consumers felt were real: The zippers on cartons did not zip; the opening features on cans often cut; reclosure features did not protect the product; containers were difficult to open; products spilled and product removal was messy; contents could not be used up completely.

But we were the pragmatists of industry! We could take every negative experience of consumers and not only explain it, but actually justify why it was appropriate. To do so, we appealed to a higher cause: line speed, labor use, capital cost, materials cost—values we judged to be of greater significance. And our goals were more lofty. Our sights were set on "return to shareholders," "being a good corporate citizen," and "providing a productive environment." The consumer's irritations were real, but the root causes of those irritations were not viewed by the packaging industry as being significant enough to address.

That was a decade ago. I now know that the consumer is right! Not only is today's package determined by the consumer, the consumer also defines the production line and machinery systems. This, at least, is the vision within the enterprising, contemporary sector of the packaging industry.

Why the change? What is so different one decade later that prompts a different perspective? The answer lies in the basics of business—how our industry defines its goals—and our recognition that the strategic plan must be supported by a strategic process. Together with goals, the strategic process dramatically alters how one focuses on issues, how change and the competitive environment are viewed, and what commitments industry will make to effect change.

In the traditional environment, the annual goal of business is to increase return by implementing the strategic plan. The difficulty with the goal of "increase return" is that, for most employees, it is difficult to personalize on a routine basis. Is the company goal of "return" really a tangible part of an individual's planning task, projects, and communication for the day?

Goal issues aside, the problem with traditional strategic plans is that their scope is based upon the past and present. Plans are, in effect, designed to yield incremental changes based on today's knowledge and experiences. The focus is on more. While that is certainly appropriate for growing a business incrementally, the difficulty with it as a strategic plan is that it fundamentally relies upon competitors remaining constant in their own planning and product development.

Attempting to achieve goals through incremental change has routinely led us to a lack of fulfillment in our execution. How are evaluations of change and innovation addressed? Often the path taken first is to route inquiries regarding packaging innovation through purchasing managers. This has led to a constriction of packaging and engineering departments, among others. Purchasing managers query vendors about "change" and "innovation" and adopt positions and directions based upon what they are told—not upon what they have gone out and invested time in to learn first hand.

There are several reasons why this process is inadequate:

- *Companies are focusing on fewer vendors to supply packaging materials and systems.* Their rationale is that a greater concentration of purchasing dollars will have a positive impact on supplier commitment and price. This, of course, is accurate, but the costs to achieve these benefits is high. The packaging industry is full of manufacturers and distributors with innovative concepts and products who are unable to get an audience with a potential customer simply because they are not on the "preferred supplier" list. Optimizing cost carries with it a decrease in competitive differentiation—and that is a high cost indeed.

- *Vendors evaluate a request for innovation against the backdrop of promoting their own company resources.* As a result, they do not always objectively address client goals. All too often, the path taken is to evaluate opportunities for innovation on the basis of what can be done with the assets now employed—with the equipment and resources presently available to the vendor.

- *Evaluations by vendors are "free."* Why are vendors so often chosen to evaluate innovation? Because they are perceived as free. Vendors will protect and advance their business with services, developments, and packaging exploration that they perceive will enhance their own prospects. Also, it takes virtually no commitment from users of packaging materials and systems to ask key vendors for analysis and development. Is this the best approach? The answer is "yes" if the lowest development cost is the yardstick for measuring innovative development. However the answer is "no" if objective, thoughtful, quality responses are sought.

- *The task under discussion is most likely to be of short-term duration—too short for a fair, accurate, and objective evaluation of true growth opportunities.*

Frozen juice concentrate packaging fits the model outlined above. The annual output of frozen juice in the United States is huge—over one billion cans. The market for frozen juice concentrate is stable with little apparent growth projected. There are only a few processors of national brands (i.e., Tropicana and Minute Maid).

The challenge in the frozen juice business in the last decade has been to make frozen juice concentrate packaging cheaper, not to increase consumer use. Thus the packaging industry's goals were to "Make the composite cans faster," "Use one or two wraps of paper instead of three," "Use a lighter gauge metal end," and "Decrease spoilage with the plastic opening feature." These are all responses to incremental growth objectives and strategic plans; but they do not reflect consumer needs. They do not address the changes in lifestyle that make the time needed to thaw frozen juice unacceptable, whether it be to run warm water over the container before opening or to set it out the night before. The consumer wants this "nuisance use" of the product to be eliminated.

These are the issues, and frozen juice concentrate is an example of the packaging industry acknowledging and understanding the irritations of consumers, but nonetheless reverting to a supposedly higher order of need (e.g., making the package cheaper).

It is difficult for companies to become more enlightened because the focus is directed elsewhere. Marketing and sales managers are the first level of planning and execution.

They identify the market for a company's product, establish the customers who will use the product and determine the price at which the product will be traded. Once this is accomplished, they set their sights on preserving that business against competitive threats. Price is the bargaining chip most commonly employed between buyers and sellers, and multiyear contracts are utilized to fulfill each party's financial desires.

The goal of being a low-cost producer is a good one. But the traditional approach to being a low-cost producer spearheaded by sales managers arguing that with lower costs, more could be sold, led to the spending of capital monies on being "the biggest" in order to have the lowest operation costs.

The largest printing presses and molding machines were often purchased with the confidence that they provided low-cost manufacturing. Since fewer companies could afford the more costly large presses and machines, an added competitive advantage accrued to investors in "large" technology. However, these larger presses took longer to set up. So a business strategy was adopted based on the idea that one makes money while the press is running, not during make-ready.

As a result, very long runs of common, repeatable, national account products became the focus. "Distinction" in packaging eroded; fitting on those large, high-speed presses was what mattered most.

But don't misunderstand. There is absolutely no question that incremental growth and being a low-cost producer are critically important goals. I argue, though, that they cannot and should not constitute the only goals and plans of a company.

The issue is whether sufficient attention has been paid to the consumer. Were the members of Mona Doyle's consumer panels truly tilting at windmills? No, not really. It has just taken the packaging and packaged goods industries a long time to give proper recognition to consumer needs.

A forerunner of this change in industry perspective was Double H Plastics in Warminster, Pennsylvania. In the mid-1980s, Double H acquired technology called Versaform, which inserted paperboard discs into injection molding machines and then formed molten plastic around the discs' edges. The principal value of using paperboard is its printability, while the plastic supplied the functional aspects of lids. The Versaform technology had floundered for ten years prior to the Double H acquisition, partially due to an initial emphasis on using eight-, sixteen-, or even thirty-two cavity molds—again, an example of "larger means cheaper" thinking.

Double H changed the scope of the technology by using a series of one-up molds on small molding machines. By setting up a row of these machines with a take-away conveyor running beneath them, Double H was able to substantially reduce handling costs. The less expensive molds allowed Double H's clients to grow as consumer demand progressed rather than incur substantial up-front costs, which substantially reduced risk.

As early as the late 1970s, Quaker Oats was looking for a replacement for their internally manufactured formed paperboard lid on their round oats canisters. Their old lid was opened with a string. However, the incidence of string failure in opening was high, and consumers continuously complained. Quaker Oats had known about Versaform for many years prior to its acquisition by Double H and had considered it viable technology, but not at the then prevailing price. The lower-cost manufacturing strategy and "pay as you go" versatility allowed Quaker Oats to invest in the technology, and today the Versaform closure and overcap—with its colorful graphics and useful recipes—can be found on every Quaker Oats canister.

I suggested earlier that how our industry sets goals and plans has traditionally had an impact on whether or not we acknowledged consumer interests. A classic example may be found in the evolution of the new rounded-rectangle ice cream package being used by Good Humor/Breyers Ice Cream (see page 66).

For years, research has consistently shown that consumers dislike square, paperboard ice cream cartons. Folding cartons do not open easily, are messy to use, do not reclose securely, and do not effectively protect the unused product. Consumers rushed to the alternative, a more costly cylindrical straight-wall paperboard container. It became standard practice for ice cream marketers to increase pricing by ten cents per half gallon while at the same time increasing sales with the round canister. As a result, the round container was used prolifically—so much so that no dairy could claim packaging as a distinctive and differentiating merchandising factor. A "signature" package is an integral part of Good Humor/Breyers' strategic process based upon differentiation. The trademark SpaceSaver package provides that uniqueness. The package employs the consumer-friendly features of the round canister, while providing the cube and material cost advantages of the square paperboard carton.

Does the rounded-rectangle container sell more ice cream than the other packaging choices? Good Humor/Breyers does not know the answer

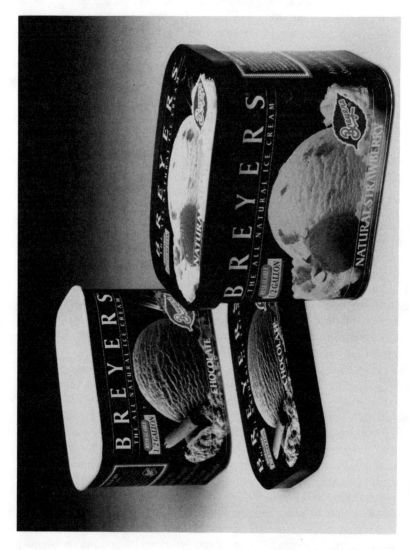

The rounded rectangle ice cream package being used by Good Humor/Breyers Ice Cream.

because, in addition to the package change, the company simultaneously initiated heavy and continuous price promotions and introduced a wide range of new flavors, different butterfat contents, and a new frozen yogurt line. Which of these factors caused a surge in Breyers' sales? Although consumers have clearly voiced their approval of the new package in a variety of consumer research studies, the answer probably resides in all of the changes.

But let's return to the question of goals. When they launched their ice cream campaign, Good Humor/Breyers did not have as their goal short-term profits or return to shareholders. Rather, they clearly recognized the constant flux of the ice cream industry, led by regional brands vying for sales with price and advertising promotions and periodic flavor changes. Good Humor/Breyers, therefore, defined its goal as becoming the market leader in all major markets nationwide.

Now that is a goal one can get behind, because it defines objectives within a structure of competitors and competitive action. Employees can come to work each day addressing their specific job responsibilities with the enthusiasm and verve of being the best. And profits will come to a much larger extent by establishing that market position first. It was this kind of proper goal setting by Good Humor/Breyers that allowed consumers to be heard—those consumers who had repeatedly voiced their preference for the rounded-rectangle container over the traditional round canister by as much as a nine-to-one margin.

The story of Nile Spice Foods and their decision to self-manufacture their microwavable soup cups offers more compelling reasons for taking a fresh look at objectives and managing one's business with a strategic process.

In 1989, Nile Spice Foods began making a line of flavorful specialty soups that could be easily reconstituted with boiling water or in the microwave oven. In a short time, Nile Spice was marketing over twenty million paperboard cups of soup, which could be found in lunch boxes and sacks wherever the company could establish distribution. The cups were supplied preformed by a prominent cup converter, who used forming equipment purchased from Paper Machinery Corporation (PMC).

As Nile Spice's business grew, the company sought to reduce packaging costs as a means of providing the funds needed to reach their defined goal of expanding to nationwide distribution. A local cup converter with a fledgling new operation approached Nile Spice and offered to supply preformed cups at prices 25% below what Nile Spice had been paying. Curious about cost/price relationships in container forming, Nile Spice approached Paper Machinery Corporation with the purpose of evaluating self-manufacture of containers. (PMC will sell directly to users somewhat reluctantly since it is the cup-forming machinery manufacturer of choice for all of the major cup converters. Selling machinery to users—in this case Nile Spice—has the appearance of PMC being in competition with its customers, a position it is pledged to avoid wherever possible.) PMC provided a machinery quote to Nile Spice, but at the same time called the current cup supplier as well as another cup converter to Nile Spice to promote a different plan. The proposal was that a converter place a PMC machine in the Nile Spice plant, operate the machine for Nile Spice, and continue to supply printed sidewalls for the container-forming process. This could be done at a price competitive with, if not lower than, the "low-ball" bid by the fledgling converter trying to fill available capacity. Both converters declined. Nile Spice recognized the cost effectiveness of self-manufacture and purchased a container-forming machine from PMC.

Self-manufacture of containers proved to be over 45% less costly to Nile Spice than buying preformed cups. The machine was paid for in less than one year. The savings were put into expanded distribution and product development. The business tripled and Nile Spice ordered a second machine from PMC to handle the expanded capacity. Nile Spice's success stimulated Quaker Oats to purchase the company in 1995 and fold the Nile Spice business into their Golden Grain pasta, rice, and soup business.

Nile Spice was, of course, pleased with the results of their decision process. Their objective was achieved. But what were the objectives and strategy of the converters—those companies that form containers as an integral part of their business? A converter with an objective stated in terms of "profits" and "return on assets" views the increase in industry capacity caused by self-manufacture, and the price to be paid for new forming machines, as inconsistent with these goals. Likewise, if a converter's strategic plan includes "decrease costs," "increase capacity utilization," "increase speeds and output," and "decrease spoilage," the reluctance on the part of that converter to "manage" a customer's in-house manufacture of containers is understandable.

But what happens to the decision process if the business objective is something else? What if the objective is "We will be the best," "We will be the supplier of choice," "We will have the largest market share," or "We will grow faster than any of our competitors"? What happens when we move beyond strategic plan and employ strategic process and that process states we will grow the business by developing products and markets that satisfy our various customer constituents—including consumers?

What happens is that we soon discover that this supposedly mature and stable paper cup and container business can grow. Had a contemporary, entrepreneurial approach to goals and a strategic process been taken with the Nile Spice opportunity, a cup converter would have stepped forward with a plan to manufacture containers in the Nile Spice plant. Keep in mind that it was never Nile Spice's desire or intention to manufacture containers. Their goal was simply to provide optimum value to their customers by increasing flavor choices and expanding distribution.

It is encouraging to note that situations in the marketplace suggest that the manufacturing companies with traditional goals and plans are becoming more enlightened. For example, one year after Nile Spice started forming containers in-house, Sweetheart Packaging purchased new machines from Paper Machinery Corporation to run half-gallon, tapered ice cream tubs at Blue Bell Creameries, the major ice cream manufacturer in the Southwest.

Of course, there were economic advantages to Blue Bell in this arrangement. But this marketing approach also promotes the strategic process of Sweetheart Packaging—the process of becoming the prominent packaging supply firm for the frozen dessert industry. It would have been infinitely easier for Sweetheart to increase their packaging supply business with Blue Bell incrementally, growing as Blue Bell's business grew. But Sweetheart understood that their competitive environment is dynamic, that any one of their competitors may convince Blue Bell to change their package, change their terms of sale, even change their source of supply. Sweetheart's use of the strategic process will, over time, be much more significant than a strategic plan focused on incremental growth.

I have argued that business objectives must have a focus that exceeds incremental change from current business. Certainly Double H Plastics did this with their evolutionary approach to the injection molding of composite lids. And Nile Spice committed to in-plant container forming as a means of achieving their business objectives.

I further argue that concentrating on the strategic process (i.e., being the biggest, being the best, pursuing greatness) is an excellent means of motivating all employees to optimize performance. Breyers has used its new rounded-rectangular half-gallon ice cream package to create excitement among its employees as it seeks to dominate the frozen dessert business. Sweetheart Packaging used a different concept: manufacturing containers in-house for customers, an approach that marries the expertise of container converters with the economies of in-house forming, and aims at positioning Sweetheart as the best resource for packaging.

These examples support the principles of expanded objectives and the strategic process. But this discussion started with the premise that the consumer defines new packages and the machinery that supports those packages. The package developed for Breyers certainly supports that premise, and consumers have also appreciated that Quaker Oats replaced failing string-opening lids on the oats cylinder with the EZO closure from Double H Plastics. The conclusion: It is consumer demand that can bring focus and perspective in pursuing business objectives and the strategic process.

I noted earlier that cheaper frozen juice cans has been the incremental objective of the large juice market segment. An enlightened objective would be to grow the market—or one's market share—by changing how the product is valued by the consumer. In today's fast-paced world, with eating often done on the run, it makes sense to eliminate the thawing time required to mix a pitcher of juice. How can this be accomplished? By replacing the metal ends of the can with ones formed from paperboard. The functionality remains the same, but one usage feature is different: A can free of metal is perceived by consumers as one that can be safely put in the microwave oven for quick thawing. It is not technology that is limiting this change. Rather, it is the juice manufacturers and the can suppliers who have been content to measure progress against incremental growth objectives. The consumer has taken a position about ease of use and has increasingly chosen to buy fresh or reconstituted juice in more convenient packaging, namely, gable top paperboard containers—many with expensive plastic fitments.

Cereal is another market in which the consumer position on packaging is clearly known. The bags inside the boxes tear and cannot be reclosed easily for product preservation. Stale product at the bottom of the box is often thrown away. Children spill the cereal when pouring or by tipping over an unstable box. In fact, the bag-in-box cereal package is amazingly unfriendly to consumers, given the relatively high cost of cereal and its high incidence of use.

However, the objectives of the major cereal producers do not reflect a consideration of consumer use or convenience. Instead, they are incremental objectives of decreasing packaging cost or increasing package value without increasing cost. The latter led to the development of high-gloss, scuff-resistant lacquers and coatings which enhance package appearance. So we are resigned to hear carton converters extol their competitive value on the basis of having the most recently purchased or largest or fastest press on which to run cereal containers.

But "achieving greatness" requires greater differentiation from competitors. It requires making a unique contribution. And, if it is to be sustained, the differentiation must satisfy the consumer—the ultimate buyer. How does a packaging supply company and a cereal manufacturer achieve this uniqueness? Not with a larger, newer press, but rather with a resealable carton that preserves the product, that has no inner bag, that is easy to open and reclose, and that has costs properly reflective of its value.

Again, the technology exists. The system to form the carton described above is being developed at Paper Machinery Corporation. But this system will only be evaluated properly for cereal packaging when some enterprising company steps forward to set objectives oriented to the consumer—objectives that are not incrementally based on existing package line production or marketing of product—and when that same company is attuned to the strategic process through a dedication to "being the best."

The same is true of frozen entrees. How many consumers have precariously handled eat-in paperboard trays when removing them from microwave ovens? Consumers use two hands when removing frozen entrees from the oven because they do not trust the package integrity or strength. Some even immediately put the entree on a dinner plate—which defeats the "no clean-up" feature of eat-in trays. This need not be. Technology exists within the paper cup industry to make economical, disposable, rigid trays that satisfy consumer needs. Indeed, many industries and product segments would benefit from adding expanded objectives and the strategic process to their corporate cultures.

Although we have cited examples here of enlightened companies serving specific consumer needs, the positions expressed here suggest that, in general, industry is at fault, that it is not prepared to step up to the tougher, nonincremental business objectives. Industry is not generally committed to, nor does it heartily endorse, the less definable, less tangible strategic process.

But the truth is, the consumer is not consistent. The consumer is easily influenced by a barrage of conflicting messages, and is constantly forced to balance information and take positions. "Paper is more recyclable than plastic"—or is it the reverse? "Packaging is filling up landfills'—or is it less harmful than other materials? Packaging's main function—to protect and transport product—is evaluated against questions of overpackaging: "Am I paying for product—or am I paying for packaging?"

In the 1980s, these concerns led to Budget Gourmet's successful marketing of a microwavable, eat-in tray. Package cost was minimized, convenience was optimized, kitchen clean-up was eliminated, the package was perceived as recyclable, and the consumer believed that most of the purchasing dollar was spent on product, not packaging. Yet I recently watched a consumer panel conclude that although an eat-in tray is still desirable, it should be inside a carton and not be the merchandising and distribution package itself.

Any machinery company—Paper Machinery Corporation included—must spend several thousands of hours engineering customized machinery to form packaging the consumer demands. Quality machinery manufacturers are well equipped to evaluate and measure the risks of achieving the designs, fabrications and assemblies that will culminate in commercial packaging machinery. But these same companies have no ability to assess the risks associated with consumer behavior and attitudes. Some companies want badly to find the products that will differentiate them from the competition. There absolutely must be a better means of pursuing market satisfaction than relying upon incremental growth based upon selling to a marketplace that one's press is 50" versus a competitor's press size of 40".

So the risk must be reduced to enable more companies to follow an enlightened path of management. And that means that consumers must understand the role of packaging within their lifestyles, they must be able to articulate their attitudes, and they must be steadfast in their beliefs. Just a few years ago the consumer was relentless in pushing the packaging and packaged goods companies as well as government agencies to change packaging to be "environmentally appropriate." The packaging industry has spent huge amounts of time and money to accommodate that demand. After all, at the end of the day, we are consumers, too. Today there are many of us in industry developing "next-generation" packaging and packaging systems because we continue to place "environmentally appropriate" on the "must" list of criteria. But we also are on the threshold of being deserted by the consumer—if, indeed we haven't already been. I recently heard a panel of consumers say somewhat in unison that they would pay ten cents more for convenience even though it doubled the amount of packaging. This absolutely is contrary to most consumer studies over the last several years.

"Next-generation" ice cream packaging is now being marketed by Good Humor/Breyers. The same is true of packaging for grated cheeses from Kraft, hot cereals from Quaker and soups from Knorr. But if we are going to develop "next-generation" frozen entree, cereal, and frozen concentrate packages—the tough ones—we must learn and understand the steadfast resolution of consumers' interests, particularly as they contribute to consistent purchasing decisions; ones not easily changed over time.

No, the consumer is not "tilting at windmills." We in industry are poised to pay attention and respond to the directions of the wind. However, the winds must be consistent and constant in the production of energy. But the winds are not always constant. Producers who accept this risk and reach beyond incremental goals will participate in a strategic process that is directed toward achieving greatness.

GERALD P. MEIER

Mr. Meier is the vice-president, Packaging for Paper Machinery Corporation in Milwaukee. He joined PMC in 1987 with the objective of starting up a packaging business within PMC, a company known principally as the leading worldwide manufacturer of machines to form paper cups. The goal was to develop a packaging business that equated to at least one-third of PMC's total business. That goal was reached in 1994. Mr. Meier was awarded the Worldstar Packaging Award and the James River Gold Key Award in 1992 for development of the Breyers half-gallon ice cream container. Mr. Meier has previous experience with International Paper in New York as manager, Business Development for the consumer Product Group and over fifteen years with Container Corporation of America in various sales and marketing management responsibilities.

Mr. Meier is a liberal arts graduate of the University of Wisconsin–Madison. He is a member of the Institute of Packaging Professionals and the Packaging Machinery Manufacturers Institute.

VIII

What Magazines Write about Packaging: No News May Be Good News

As of the spring of 1995, food packaging per se is not a subject of much interest to magazine journalists. Perhaps, this should come as no surprise. Whereas the electronic press (television and radio), weekly magazines and newspapers generally report and analyze information that becomes important almost overnight, monthlies more often develop articles for other reader purposes. These publications usually focus on:

- how to solve problems
- how to evaluate the pros and cons on controversial issues
- how to apply new research findings in the marketplace

One might surmise, therefore, that packaging is not a consumer problem, nor a very controversial topic. Although research continues to improve packaging technology, consumers are not required to learn much in order to integrate such changes into their purchasing decisions. All of these explanations bode well for packaging professionals.

The Few Exceptions

Occasionally packaging articles do appear. The closest thing to a hot topic occurred when baby food manufacturers downsized their jars without reducing their prices. The plastics-versus-paper topic continues to play in magazines such as *E, The Environmental Magazine* and was often written about in *Garbage*, a magazine that is no longer published.

Several years ago, a magazine wrote about how package designers use color to create a product image and stimulate trial purchase. Also, loss of

75

vitamin activity due to light exposure in products with clear packaging has been casually mentioned in a few magazines.

Canadian Living, as part of a larger piece about arthritis, wrote a half page in late 1994 about the packaging dilemma due to the need to balance ease-of-opening against preventing product leakage, allowing better container stacking, and deterring tampering. What may be an emerging story, covered in the April 1995 issue of *Child*, is the possibility that talking cereal packages and temperature sensors for microwavable foods might soon be coming to grocery stores.

Interesting Omissions

Compared to the trade press and food technology journals, the absence of certain topics in consumer magazines merits consideration. Sous vide packaging has received relatively minor coverage, nor have journalists yet given much attention to resealable closures, office-microwavable lunch items, cheese-and-cracker type combined packaging, or child-size-portion products such as yogurts and desserts.

Magazines frequently publish charts comparing brand-name products in the same category. Columns include nutrition information (usually fat, percent of calories from fat, calories and sometimes sodium), price and results of taste-tests conducted by editorial staff. However, evaluation of packaging benefits versus problems has, to my recollection, never been included as a product attribute variable in such brand-comparison articles.

Buried in Bigger Stories

Food processing technologies are news. Many magazines have dealt in some way with food irradiation but only rarely have editors mentioned the pink-flower logo on these products' packages. Similarly, genetic engineering articles escalated when the Food and Drug Administration (FDA) approved the use of bovine somatotropin and then the Calgene tomato.

Lengthier articles on this topic discuss some people's desire that the packaging of such products be labeled accordingly, but such information is simply included with other consumer concerns about this type of biotechnology.

Food safety is another important issue for magazine journalists. The number of articles on this topic increased significantly after deaths occurred due to *E. coli*-contaminated hamburger. Editors often warn consumers about the danger of undercooked ground meat when eating out and publish advice from the Department of Agriculture (USDA) about safe handling and preparation of foods in the home. But again, the important observation is the omission—magazines are not writing about safe-handling labeling mandated by USDA. Perhaps journalists think these messages are self-explanatory, or maybe it is too difficult to make this packaging change interesting to readers.

Package Labeling

For several years before the revised FDA regulations were published, the problems with potentially misleading label information were covered extensively by consumer magazines along with advice about how to avoid misinterpreting certain "health claims." During 1994 and 1995, most magazines wrote something about the new Nutrition Facts panel on food packages. In general, the message is positive and educational. Some articles simply print sample label artwork from FDA; others develop their own graphics to emphasize specific points made in the editorial content.

Journalists are now moving beyond the basics regarding Daily Values and Percent Recommended Daily Intakes. More detailed information, such as the significance of sugar values and how to use percent of calories from fat, is being addressed.

In early 1995, *Self* reported that the Federal Trade Commission (FTC) had charged that Häagen-Dazs advertisements made unjustifiable claims for the caloric value of three frozen yogurts and some of its "low-fat" products. When and if other companies are found to be out of compliance with either the FDA or FTC regulations, magazines will no doubt make this widely known to their readers.

What's in the Crystal Ball?

Some magazines, especially January issues, write articles about what might happen in the new year—or even in the next decade. Editorial style varies from serious science reporting to whimsical humor or political satire. In this context, here's what might be predicted, based on magazine forecasts of changes in how food will be marketed in the future and evolving opportunities for packaging experts.

- The increase in consumer comfort with electronic technology will open new opportunities for purchasing many foods without having to walk the aisles of a grocery store; packaging changes may facilitate this transition by solving delivery-related problems that currently restrict the types of foods that can be shipped directly from producers to consumers' homes.

- As health-conscious consumers progress from knowing they should eat more fruits and vegetables to trying to do so, packaging technology may play a critical role in providing improved quality and better shelf life (at home as well as in the grocery store) for fresh produce.
- As more people retain their health and vigor to older ages, packaging innovations might be the key to developing creative products for this new group of eat-on-the-go consumers.
- As environmental concerns and convenience demands merge, the ideal package will stable at room temperature and be edible after it has been cooked in the microwave.

Tomorrow's consumer news about packaging is, no doubt, already in today's research and development laboratories. Consumers await what's in store for us all.

KRISTEN McNUTT

Kristen McNutt edits and Consumer Choices, Inc. publishes the *Consumer Magazines Digest: Nutrition and Food-Related Health Topics*. This monthly publication summarizes for subscribers articles in more than fifty current consumer magazines. This chapter is based on Dr. McNutt's assessment of packaging-related articles written in the popular press over the last seven years.

IX

SHELDON B. SOSNA

A Retail Advertiser
Looks at Packaging

It's 1996. Do you know where your package is?

Because design is now not only dependent on what's inside the package but where and how the package will be sold to buyers. Sure, packaging must do all of the nifty things outlined so far in this book. It must attract the shopper, say something nice (even truthful!) about what's inside the package, be as kind as possible to the environment, adhere to the corporate philosophy behind the brand and be as innovative as possible. And so on and so on.

But there are a small handful of very down-to-earth, pragmatic tasks that successful packaging must accomplish on its journey from manufacturer to kitchen. Often these are overlooked as marketers focus on the big picture, the broad strokes, the marketing strategy and the bottom-line.

The fact is that, very often, the success of a brand depends on another important factor: How well the package can negotiate its way from the warehouse to the cash register. What happens in the store can be as important as anything else that happens to the product.

At the Dock

If you haven't followed a case of something from the factory to the warehouse to the store and, finally, to the home, you may not know everything you should know about the package you are designing/producing/using.

What about the shipping cases in which the product is packed and delivered to the store? If they don't clearly spell out the kind of product that is inside the case, it can get lost in the warehouse, on the shipping or receiving docks, or in the back room of the store. Only if you've seen a case of frozen orange juice melting in the sun because it didn't say "Keep Frozen" prominently enough, will you be able to understand the importance of having cases that are easily identified as to brand, package size, and care instructions.

The guys and gals who work on forklifts and trucks in warehouses and retail store back rooms are too busy to spend time studying your cases. Your products will get handled as they *look* they should be handled. So don't make the case for a perishable product look like a case of canned soup; don't make the case for a fragile product look like a case of paper towels; don't make the case for a new product look like it's the case of a product that the people who pick and pack are familiar with—especially if it's a different kind of product!

Although they're not directly related to package design, display materials can be important to the success of a product label, especially a new one. Very often, new products are shipped along with expensive display materials: shelf posters, self-display cases, store signs, and shelf talkers. These materials should come to the warehouse packaged with the product, not in separate shipping cases. That way, they can't get "lost" on their way to the store. When my company researched the matter of undelivered display materials several years ago, we discovered that materials shipped separately from product got lost on the way to the retail outlet much of the time. Often, they are very expensive four-color pieces that manufacturers spend big bucks on. Too bad they get used to blot up puddles on a shipping dock after a rain! Or go into the dumpster without a second thought.

Also, if the product is to be shipped in self-display cases to be assembled in the store, they must be easy to put together. Unless they are no-brainers, they are certain to be botched.

And, finally, are your cases and packages stocker-friendly? Can the kids charged with extracting the product from the shipping case and putting it up on shelves do their job easily? That means that the cases must be easy to split open without slitting the individual cartons within, that the product is easy to decant from the case, and that it is packed in such a way that it can be price marked (still necessary in many places!) and swooped from case to shelf in one easy motion.

In the Store

While most people engaged in designing packaging for consumer products spend time and effort in perfecting the appearance of those products, they must also consider where the products will be offered for sale in the store. We all hope for the best—an eye-level placement on supermarket shelves, for example. However, the fact is that only about a third of grocery store products get such advantageous positioning. The other two thirds share shelf space that is above and below that optimum.

And we have to realize, also, that in an effort to conserve shelf space, some products are often placed side-out on shelves rather than face-out.

Do these facts change the design of a product's packaging? We think so. For products that are often placed below normal eye level in the store, the top of the package is almost as important as the face. Somehow it must get people to look down at the product. For products that are sometimes relegated to top-of-the-gondola display positions, the packaging must say *look up here* to attract the shopper's eye. And to guard against invisibility in the store, even the side panels of some packaging must be made as attractive as the front panel.

In the Ads

Perhaps the most important retail consideration for many products is how often they get featured in store advertising. Even a product that gets considerable media support from manufacturers in magazines and on television and radio needs additional support in the store's own advertising. If it doesn't get featured in store circulars and ROP newspaper ads, it can't live up to its real market potential.

Take a moment to consider how art is used in most retail advertising. While some products are shown in large and very large size in the ads, the vast majority of the products featured in print, especially by mass-market stores and supermarkets, are shown in postage-stamp size. Try designing your packages to stand out in a 1-1/2 inch by 1-inch illustration. If you can't do it, get into something easier!

Several other factors affect how well a package design works in terms of its use in retail advertising.

The first is how available package artwork is for use in retail ads. Does the manufacturer make package art available to retailers? Most grocery artwork, for example, is available to supermarkets via the Kwikee books of package slicks in black and white and color. These are now issued four times a year (at no cost to the retailer) by Kwikee Systems Inc. of Peoria. In line with the increased use of computers to compose print ads, Kwikee also makes the art available on CD ROM disks, also issued four times a year for free to retailers.

A second consideration for package designers is how reproducible the package art is for print advertising. Packages that are predominantly white or pastel-colored tend not to stand out against the white paper on which most circulars and newspaper ads are printed. Therefore, they may get less play in ads whose layout designers are interested in achieving contrast. No matter how beautiful the photograph or artwork you plan to use in creating you package, if it includes delicate colors and soft-focus artwork, it will be hard to reproduce for print advertising purposes. Successful product labels are usually strong and uncomplicated.

When you use color in packages, the colors must be bold and simple. Like a Heinz ketchup bottle or a Campbell soup can. Soft and subtle doesn't cut it. Understated doesn't look good in advertising.

Also consider whether your package design relies on color to achieve its identification. If so, how can it be identified when it's printed in black and white, or in two colors (usually red and black!)? Most retail ads are still printed in this manner. So identification must rely more on shape and format than on color. Heinz ketchup and Campbell soup and Tide laundry detergent don't need color to be recognized. Your product designs must strive to achieve identity in terms of the shape of elements used rather than color. In terms of the average retailer, we all still live in a black-and-white world. And, finally, if the color of your package is not one that is easily reproducible in print (mauve, for example, or burnt umber), please try to be a little more primary.

And Into the Future

Anyone who does any shopping these days knows how quickly (alarmingly so) retail stores are changing: self-scanning in supermarkets, cashless debit card transactions, wildly increasing selection (17,000 new products in 1994!)—and every new product with a new label for shoppers to see and, hopefully, remember.

But you "ain't seen nothin' yet." The way shoppers "go to the store" next year and every year into the future is in for some major changes. You can see these changes coming already. Some authorities believe that by the turn of the century, many retailers will have only *one* sample of a product on display in their stores. When a shopper chooses a product electronically, it will magically appear at the cash register ready to take home as soon as your debit card is processed. Better yet, the product will be delivered to your home so you don't even have to carry it.

More and more shoppers will *never* go to a real store in the future that's just around the corner. Instead, they'll shop on TV shopping channels. Many retailers, reflecting what they've learned from Lands' End and the Sharper Image in recent years, will publish their SKUs in catalogs and take orders over the phone.

Within a few years, you will be designing labels and packaging that must look good on the Internet. Maybe you already are; shopping on Compu-Serve, Prodigy and America Online is already a multimillion dollar business. As computers get more user-friendly and we all get more computer-compatible, people will be as comfortable buying clothes and meat and hardware online as they are running down to the mall today.

So the packaging you are designing for tomorrow's marketplace must be suited to the new retailing landscape that is developing. No longer can you expect products to sit impassively on shelves and in showrooms. You've got to plan for how they'll work when shoppers are buying them in sales venues that are still in the realm of science fiction.

If you haven't been thinking about where and how the package you design will be bought by shoppers, you ought to start.

SHELDON B. SOSNA

Shel Sosna's advertising career spans both Madison Avenue and Main Street. In his Madison Avenue days, Sosna was an executive with such well-known advertising agencies as Leo Burnett, Needham, Harper & Steers and SSC&B (now Lintas) and was responsible for devising advertising and marketing strategies for such famous names as Naturalizer, Marlboro, Beefeater, P&G, Kellogg, Dr. Pepper, Pillsbury, and many others. In addition, Sosna has spent over twenty years on retailing's Main Street, working as VP Marketing for major supermarket chains and then as head of his own retail advertising consulting firm.

Today, Sosna is considered one of retailing's outstanding authorities on advertising. He publishes a monthly newsletter on supermarket advertising and writes monthly columns on retail advertising and marketing for *Grocery Marketing Magazine*. His company supervises advertising and promotion activities for a number of retailers, and he is a frequent speaker at industry gatherings. He has presented his popular "How to Win with Advertising" seminars to thousands of retailers and their advertising people at FMI, NGA, IGA, NAWGA, and at many state and regional grocers' association meetings.

Sosna has degrees in marketing from Northwestern University and the University of Chicago. He has written several books about retail advertising. He lives with his wife, two cats, and a dog in Durham, North Carolina, and shops at every supermarket in town.

Package Power

Among packaged goods marketers it is axiomatic that the package IS the product in consumer minds. Together with the brand name, it carries the history of a consumer's experience with the brand, and the consumer's rational and emotional response to that brand. Package design, of course, goes beyond providing cues to a shopper's mental file: It also can provide a competitive edge, tilting brand choice at the point of purchase. It is difficult to overstate the role of packaging in building successful brands.

Of all elements in the marketing mix, including the product itself, packaging is the most complex. The following are the major functions of packaging:

1. *Protect contents.* This includes maintaining product integrity, tamper-proofing and child-proofing where necessary.
2. *Promote ease of usage.* This includes dispenser mechanisms, single-usage inner packs, and resealability.
3. *Ensure correctness of usage.* Packages must display legal copy, directions for usage, ingredient lists, name of the manufacturer, bar codes, and dating (where necessary).
4. *Provide "instant recognition"*—through brand-name display and unique design. This function is critical to both in-store and at-home situations.
5. *Permit on-pack promotional displays.* Consumer promotions carried on the package often tilt brand choice at the moment of purchase.
6. *Display competitive information.* Nutritional advantages, effec-

tiveness data, ounces and potency, ease of usage and other comparative data often add to package power.

7. *Allow line extensions.* A good design provides for instant recognition
   across a range of segmentations.
8. *Conserve space*—both on the store's shelf and the home shelf.
9. *Conserve resources.* Many consumers react unfavorably to excessive
   packaging, such as a package within a package, and to slack fill that
   delivers only unneeded empty space.
10. *Recognize increased shopping by men and teenagers.* The
    "housewife" no longer is the sole buying agent for the family.
    Designs and package elements must be effective on a range of users
    as well as shoppers. (As with kid cereals.)

You will note there are contradictions within this list of functions. For example, "instant recognition" would be easier to achieve with larger front
panels and/or unusual package shapes, but the need to conserve space on
store and home shelves cancels this route to package success most of the
time. Unusual shapes don't ship or stack well. Large front panels defeat the
retailer's goal of high return on shelf space.

Any package design is a series of compromises within parameters set by
the product, the category and competitive brands. For example, a box of
aspirin must relate to the pain-killer category, yet be uniquely Bayer or
Tylenol. Category similarities must be respected, for they act as cues to
users. Salad dressing packages immediately communicate salad dressings.
Cereals are unmistakably cereals. Photo film is photo film. Category similarities provide instant positioning for any and all brands. This is akin to
looking at a baseball diamond: The playing field itself identifies the game
being played. Football is a different game with its own rules.

Individual brands are like individual baseball teams: They all play the
same game but are unique each to itself. The New York Yankees and the
New York Mets have different personalities and attract different followings.
So it is with brands.

Package power requires both category positioning and brand uniqueness. Package research is critical to both functions. Certain bottle shapes
have come to be associated with different kinds of alcoholic beverages,
making it easy to confuse shoppers with a bottle that signals the wrong
category.

Colors, shapes and design elements all must contribute to brand positioning. An expensive perfume cannot succeed in a bottle that signals
"cheap."

Packages must change with the times. A box of Tide laundry detergent today must accommodate Liquid Tide, Concentrated Tide and other variations. All forms must be unmistakably Tide, yet must signal their respective segments.

Package designs on well-known brands cannot undergo drastic changes, for consumers immediately translate abrupt design shifts into a change in the PRODUCT inside the package. This can have a negative impact on brand sales and shares. A well-known package designer was asked how he was going to change a leading brand. His response was "VERY CAREFULLY." The leading brand of beer in the U.S.A.—Budweiser—has had design changes over the years so subtle as to go unnoticed. Yet over time, the Bud label has changed appreciably.

Loyal brand users feel that a "contract" exists between themselves and brand marketers that no significant changes in product, package or pricing can be made without their consent—witness the uproar over New Coke.

Packages carry the brand's traditions. Coca-Cola has returned to the contour bottle because that design triggers favorable consumer responses, especially emotional responses. But this does not mean that packaging cannot be modernized to fit new audiences and new needs. For example, Vaseline Intensive Care is a "medicated" extension of traditional Vaseline. A package change sometimes is necessary to signal product improvements.

Choosing a package design is perhaps the most critical step in marketing a NEW brand. There are about 24,000 stock keeping units (SKUs) on display in the average U.S.A. supermarket today—and nearly that many in large super drugstores. The shopper spends roughly thirty minutes in one of those stores, passing roughly 1,000 SKUs per minute. Obviously, any new brand must have some ability to catch the shopper's attention, signal its category, and deliver a promise of something new and better. Most package designs for both old and new brands today work hard at standing out in category displays. With new brands, failure to have eye appeal from some distance is fatal.

But failure to "sell" the brand with on-package copy also is fatal. Today's shoppers take time to read package copy, both in the store and at home. While this may be only one package per shopping trip, over time today's shoppers become surprisingly knowledgeable about what the package has to say of concern to them. Some packages such as dry breakfast cereals go directly onto the table with the meal, so package copy has a great opportunity to "sell."

It can be said with some assurance that packages today must be better integrated with the brand's advertising for two reasons.

First is to provide a better trigger to the shopper's mental file, and especially to its emotional content. A brand name alone often is not enough to cue full brand meanings built into brand advertising.

The second is to defeat cloning by store brands, which today are imitating closely the designs of brand leaders. It is doubtful that legal suits can block a high degree of cloning, for all brands in a category tend to have similarities in design. Where can a line be drawn so long as the brand name is there and there is no consumer confusion?

On-pack use of key elements from brand advertising can help defeat the impact of cloning: A Pillsbury doughboy cannot be cloned, and the on-pack doughboy triggers the shopper's mental file, especially in the "finger-poke-in-the-stomach," which denotes warmth and freshness. Similarly, there is no way to knock off the Energizer Bunny. Or the Marlboro cowboy.

Integration of brand advertising and packaging has three components:

1. Advertising campaigns must create and imprint the symbols and words that can be transferred to packages.
2. Package designers must incorporate advertising elements into their design.
3. Brand marketers must insist that both the above get done. Too many brand marketers seem willing to include short-run promotional copy on packages, but make no attempt to include key advertising cues.

The evidence seems clear: On-package advertising provides a competitive advantage. Brand names alone cannot trigger the full set of brand meanings in consumer mental files for most brands.

Globalization of brands greatly complicates package design, for colors, shapes and words convey different meanings when the brand crosses cultural borders. Many brands historically were positioned differently country by country, and often had local package designs. With today's cross-border travel, cross-border media and cross-border marketing, these purely local positions often must be brought together, and the impact on local package designs can be enormous. Fortunately, the development of global package designers is mitigating this problem.

Although this chapter is aimed at the packaged goods industry, much of what has been said applies also to the service industry. McDonald's is a model of how to package a service business. So is overnight mail by the U.S. Postal Service.

Consumers use brands to organize both their shopping and their individuality. Packages and brand names are inseparable. Name a known brand and the mind pictures the package. *Package power means brand power.*

CHARLES A. MITTELSTADT

Chuck Mittelstadt is the advertising planner and strategist who founded and directed The Interpublic Group's Center for Advertising Services. The Interpublic Group is one of the world's largest advertising holding companies. CAS provides research and account planning support to Interpublic agencies and their clients worldwide. In semi-retirement, Chuck operates as a worldwide marketing consultant from a base in Tarrytown, New York.

HERBERT M. MEYERS, FPDC

The Role of Design

It is no accident that the subject of this chapter, the role of design, follows ten chapters that have prepared us to understand the critical role of package design in marketing a brand of products. While the previous chapters discussed some of the important elements that impact the packaging of retail products, such as marketing strategy, brand-name selection, consumer perception of the product inside and the accumulative effect of the retail environment, this chapter brings us to the ultimate goal of making the product desirable to the consumer by developing packaging that looks and functions effectively at the point of sale.

The fact is, when all is said and done, it is the package design that brings together all the factors developed during the strategic and production planning. To paraphrase President Harry Truman: The buck stops here!

It is important for you, the marketer, to realize that package design cannot function in isolation.

- Package design can attract the consumer to your product by making sure that it is noticed on the shelf among competitive products.
- Package design can present your product in an attractive manner and identify its benefits to the consumer.
- Package design can make it easy to handle, dispense and store your product.
- Package design can make the difference between your product being selected by the consumer instead of a competitive brand.
- Package design can make it easier for you to distribute your product and for your customer to store and display it.

But, there is one thing package design cannot achieve: Package design cannot rescue a bad product. Package design can only enhance what is truly there and for that it can be your most effective marketing tool.

Package design is the culmination of careful planning by marketing, sales, research, product development, manufacturing and advertising. Package design organizes the many factors generated by these activities and consolidates them into a single, all-inclusive physical and visual entity that must communicate the product positioning and marketing objectives to the consumer at a single glance.

Thus, the package designer is faced with the complicated challenge of balancing the various strategic packaging components and arriving at a design solution that will communicate the marketer's strategy effectively. The package designer becomes a juggler who must keep numerous balls in the air simultaneously without dropping any of them. Let's discuss how it's done.

Package Design as a Communicator

Everything about the package plays a role in communicating the marketing strategy. The package form itself can communicate many elements that influence consumer perception.

- Ice cream in a tub communicates a different quality perception than ice cream in a carton.
- A wide, stubby beer bottle communicates a different type of beer than a tall, tapered bottle.
- A lipstick displayed on a blister card creates a different image than a lipstick in a foil carton.
- A watch placed in a velvet box creates totally different imagery and price/value perception than the same watch placed in a plastic container.
- A straight-sided wine bottle signifies wine from France or Italy, while a bell-shaped bottle identifies it as Portuguese.

In fact, it is possible to manipulate the imagery and positioning of a product by selecting a packaging form and material that will influence the consumer's perception of the product.

Such opportunities are not necessarily a blessing. It is easy to succumb to the temptation of wanting to be "different" by selecting product forms and materials that will affect consumer perception in a way that may be detrimental to your marketing strategy. This is particularly true with regard to price/value perception. If the wrong material or package form is selected for a given product, the product may be perceived by the consumer in a way not intended by the marketer. For instance, common household nails in a foil carton would very likely be the wrong choice of package forms. So would be an expensive high-tech tool in a chip-board carton or shrink filmed on a corrugated cardboard panel.

The physical appearance of the package can and should play a major role in communicating the product strategy. Unfortunately, the economics of packaging frequently lead to packaging structures that contribute little or nothing to communicating anything about the product itself, particularly in certain commodity categories (see photo below).

Milk cartons and plastic bottles are all structurally alike, and while they quickly identify the product in the dairy cabinet, they communicate nothing unique about a particular brand. All of them are, at least physically,

Look-alike packages identify certain product categories but do not contribute to brand identity or unique product features.

total look-alikes. The same goes for many other products. When several years ago, Wish-Bone brand created a unique bottle that mirrored the shape of a wish-bone, virtually all salad dressing brands followed its example with wish-bone-like configurations. Thus, for many years, the wish-bone-like bottle became a *commodity* shape for the salad dressing industry. While it identified salad dressings immediately, it neutralized the ability of various salad dressing brands to differentiate themselves, except through label graphics. Only during recent years have some salad dressings marketers tried to rectify this situation by marketing their product in different, though often not very distinctly different, bottle configurations.

The same goes for all canned products that utilize standard shapes and sizes, relying entirely on the label graphics to communicate anything about the product. Round can configurations are a foregone conclusion in the United States. Experimentation with can shapes has been virtually abandoned here, while in other countries, such as Japan, unique can shapes for such products as beer are being explored. Fruit juices have long been marketed in many different types of physical configurations in Europe, ranging from unique glass bottle shapes to square paper containers with metal ends for fruit drinks. Only recently have U.S. beverage marketers started to appreciate that unique bottle configurations can distinguish them from competitive brands and give them a special position in their industry.

Coca-Cola, having been marketed for many years in the same generic plastic bottles used by the entire soft-drink industry, recognized the benefit of being unique by introducing plastic containers for all sizes of Coca-Cola that recreate the unique glass bottle shape that once signified Coca-Cola's leadership position throughout the world. This was a courageous departure from the stock-bottle syndrome of virtually all other soft-drink marketers and achieved Coca-Cola's objective of regaining its leadership (see page 95).

Even the automotive industry, while constantly striving to produce distinctive new forms for their cars, offers its after-market products, such as oil, anti-freeze and other additives, in standard containers, capitulating to the inflexibility of container manufacturers who don't want to bother with container shapes that require production line modifications at their plastic bottle manufacturing plants for a select number of customers.

Once caught in the trap of standardization, the ability to communicate a positive image about the product becomes extremely limited and is given entirely to the packaging graphics and the words and images which appear on the package surfaces. Thus, in many cases, the packaging graphics are almost totally responsible for communicating the market strategy to the consumer.

The new plastic Coca-Cola bottle recalls the well-known equity of the glass bottle contours.

Packaging Graphics as a Communicator

The opportunities to communicate brand identity and product benefits through packaging graphics are almost limitless. The components of the surface graphics can combine to communicate numerous messages, both informative and emotional.

Informative messages on packages include:

- the brand name and logo
- product description
- flavor or variety identification
- benefit statements
- promotional messages
- usage directions
- nutritional facts (for food)
- warning or caution information (for drugs and chemicals)
- size and contents information

and whatever else will help communicate information about the marketer's product to the consumer.

But beyond pure information, the emotional aspects of packaging graphics offer a wide range of design possibilities. These include the brand identity, the styling of the logo, the typographic treatment, symbols and icons, the package colors, textures, photography, illustrations and many more. Let us review some of these.

Brand Identification

The Brand Name

If we want to isolate the *single, most important* element on a package, it would most likely be the brand name. The name of the brand is like the name of a person. It identifies the product, creates memorability, achieves an equity that enables the marketer to build recognition and loyalty among consumers and is the cornerstone on which to build and expand the strategic objectives.

For this reason, the manner in which the brand identity is presented becomes the most critically important issue on all packages. The brand's identity is its signature. It signifies to the consumer that the product is reliable and worth considering for purchase. It helps remind the consumer to find and purchase the product on return visits to the store and, if used consistently and effectively, helps build confidence in a product and any additional products that may be introduced in the market at a later date under the same brand umbrella.

Dimetapp®

Four logos that express the character of their products: Dimetapp Cough/Cold Remedies (hard working), Van Bloem Gardens flower bulbs (gardening, feminine), Chico San Rice Cakes (fun, tasty), Cooper ice-hockey equipment (speed).

It is, therefore, critical to develop brand identification that is visually unique and appropriate for your product, that communicates a positive image about your product, and that is easy to recognize and easy to remember. Just like your own personal signature is a statement of your personality, the styling of the logo or signature of the brand name are key elements to your brand's identity and are critical in communicating a desirable image. A bold logo may communicate strength, masculinity, effectiveness. A cursive logo may communicate elegance, lightness, femininity, fashion. An angled logo may indicate casualness, fun, movement, entertainment. Even the color of your brand logo signifies an attitude and may affect the consumer's perception positively or negatively (see above logos).

Because the brand identity on your packages is so critical in communicating a positive image to the consumer, it is important to keep the brand logo as constant as possible on all products that are identified by it. Whether the logo is used to identify a single product or an extensive line of products, whether it is used as the primary brand name, a sub-brand name, or an endorsement, it is usually advisable to maintain the same logo styling and proportions for all packages. Many companies do not follow this theory, however. They alter the proportions of the logo depending on the proportions of vertical or horizontal packages, adjust them on a variety of packaging shapes, or modify them in connection with various graphic treatments. They also change logo colors and sometimes even allow other package designers and printers to reinterpret the logo styling, proportions and colors as they see fit.

These are serious mistakes that should be avoided under all circumstances. Your brand identity is the most important link in your chain of communication, whether used in connection with packaging, print ads, TV commercials, sales promotion, signage or your letterhead. The identity of the marketer's brand, as represented by the brand logo, should not be modified, except under the most extenuating circumstances, such as adjustments necessary to accommodate printing restrictions on certain substrates.

Even the color of the brand logo is important and works best if kept constant. While exceptions may be considered when flavor or variety situations dictate the need for color differentiating to identify them, it is advisable to ensure that the logo is strong enough to retain its character and recognizability.

Sub-branding

When a brand consists of a single line of products a constant brand name will, in most instances, be the appropriate method of brand identification. This is particularly true in the case of commodity products that do not require product differentiation other than type of product or product use, color or flavor, and package size.

When there is a need to identify a particular product quality or describe a distinct product characteristic, a marketer often emphasizes the unique nature of such a product by means of a sub-brand name. The sub-brand name most often attempts to be unique and memorable, relating to the nature of the product or creating a perception that the consumer can identify with, such as an emotion, an experience, a place, a description, or a perceived value.

Many sub-brand names have played significant roles in the marketing of products. Macintosh QUADRA, Oldsmobile CUTLASS, Tropicana TWISTER, Ziplock GRIPPER-ZIPPER, Crest COMPLETE, Corning PYREX are a few of the numerous examples of successful sub-brands.

The styling of sub-brand names usually differs from the style of the primary brand name in order to differentiate between the two brand entities and, in many cases, to add a communicative meaning to the sub-brand.

One of the most dramatic examples of the significance of sub-branding is when Coca-Cola decided to modify its popular flavor formula to better compete with Pepsi-Cola. When this resulted in an outcry from loyal users who wanted no part of the flavor change, the well-known brand retreated to the original formula, identifying it with a single word, "CLASSIC." After the initial monumental blunder, this was a brilliant move by the beverage marketer because the consumer immediately and clearly understood that Coca-Cola CLASSIC meant the brand's return to the original formula. Thus, a single word on a package, further strengthened by its classic logo styling, was able to regain the strategic position that Coca-Cola had occupied for over seventy years (see page 100).

When, for various strategic reasons, Sudafed Cold and Cough Remedies planned to change its packaging graphics, preliminary marketing research indicated that one of the products in the product mix was not understood by the consumer. This product was identified as SUDAFED S.A., which stood for "Sustained Action," a term used internally by the product's manufacturer, Burroughs-Wellcome, to identify its longest lasting product variety. The designers recommended calling the product SUDAFED 12 HOUR because this more adequately identified the product benefits to the consumer and created an easy-to-remember brand name for the product. Again, a simple but precisely targeted use of words made a substantial improvement in communicating the product description to the potential user (see page 101).

Copy

Next in importance to brand identity on packaging are the verbal communication elements, that is, the words that appear on the packages to identify the product and provide information about the product. Because they have to be carefully structured in relation to the limited space available on the package, every single word on the package has to be precisely targeted to communicate the intended strategy.

Contrary to television commercials, with packaging copy, there is no opportunity for music and sound effects to set an emotional stage and thus assist in communicating a message to the consumer. Further, no movement or special effects are available to the package designer in creating a specific environment or mood, no voice-overs to explain a product feature, no facial expressions to help in appealing to the emotions. Everything is condensed to the minimal proportion of a package or a label. Every word and every visual detail has to be so carefully worked out and specified as to leave no doubt, at a single glance, what the product is all about.

In most instances, the time the package has to identify its contents is counted in seconds. If the package does not attract the consumer's attention and communicate instantly, the sale may be lost. For these reasons, the general rule for verbiage on the packaging is: The fewer words, the better—so long as the words are carefully and precisely targeted to clearly communicate the brand's position and product benefit.

The introduction of Coca-Cola "Classic," identifying the original taste formula, rescued the brand from near disaster.

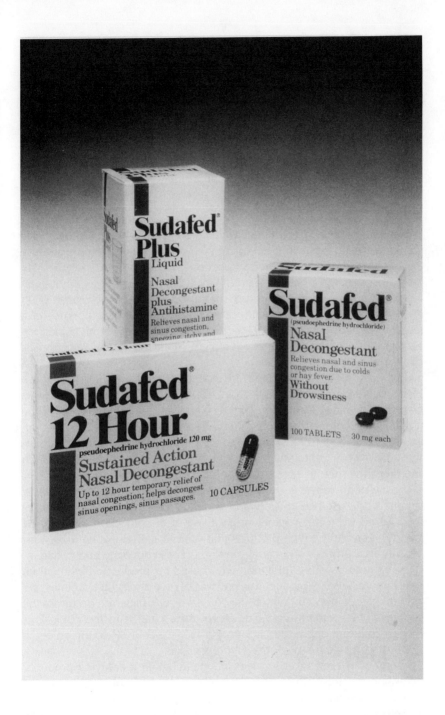

Sudafed "Plus" and Sudafed "12-Hour" clearly identify the benefits of each product.

Color

Color is another visual element that can communicate many different aspects ranging from utilitarian to emotional:

- Color can set a mood such as fun, elegance, flavor.
- Color can help create quality perception.
- Color can identify the color of the product inside the package.
- Color can assist in differentiating products, varieties and flavors.
- Color can identify a brand (Kodak, Healthy Choice).

Examples of how color influences purchase patterns are numerous. They include cereal boxes, which frequently have bright, lively colors because they are consumed in the morning, a time of day that is associated with brightness. Similarly, light or diet food most often appear in white or light-colored packages to communicate their product attributes. Frozen gourmet dinners, on the other hand, often utilize deep, rich colors to set a mood of elegance and luxury. Gray and black colors frequently appear on packages containing high-tech equipment, such as cameras, electronics, etc.

Color is a critical component in communicating product perception. Thus, pharmaceutical packages that wish to communicate an efficacious prescription image frequently feature white backgrounds instead of the bright, garish colors used by less ethically oriented OTC products. The latter seek to achieve shelf impact and are less concerned with ethical imagery. Cosmetics frequently feature colors that are associated with fashion and elegance such as pastel shades, black, and gold.

The use of foil labels is almost entirely reserved for products intended to communicate upscale imagery and is rarely, if ever, used on commodity products. On the other hand, bright touches of gold can frequently be found in connection with products that want to communicate luxury, especially in cosmetics and certain foods such as chocolate and coffee.

Use of color is more utilitarian when used for product differentiation. For example, when a product line consists of several product varieties, the only difference between which may be flavor or usage, the design of the packages often remain identical, except for a change in the color. This change could be either the background color, a color area, the colors of visual or copy elements, or the entire package.

Some colors have taken on specific meanings in identifying products. An example of this is the colors used by the soft-drink industry where red signifies Colas, green stands for Ginger Ale, yellow for Tonic Water, blue for Seltzers and so forth. If you don't believe this is true, just try to market a Ginger Ale in a red can, or a Cola in a bottle with a blue label!

The most obvious use of color is, of course, the color differences of the products themselves. For example, colored pencils, color pigments, cosmetics, printing inks, and paints are usually identified through package colors that relate to the colors of the products inside the packages, and thereby contribute to clear communication of the product differences.

Neverless, color is probably the one package design component that defies generalization more than any other. This is, in part, due to the fact that a wide palette of colors is available to the designer because of color variations in the form of lights and darks, numerous opportunities for shades and shading, and almost unlimited numbers of color combinations. Skillful use of color, therefore, becomes the most potent and most opportunistic tool available to the package designer.

Photography and Illustration

Other elements on the packages may include photography or illustrations. This is an area that requires particular and specialized skills because it involves a number of difficult disciplines. The choice of photography or the technique of rendering, the choice of models or the quality of food preparation, the arrangement of the products and the lighting of them—all of these make substantial differences in the visual impression and have a substantial influence on how the product is perceived by the consumer (see page 104).

Photography and illustrations can be used in many different ways to achieve different types of communication, such as:

- *product differences*—to differentiate among several types within a product line by simply showing the differences in their appearance, such as different vegetables, a variety of toys, various types of household goods, etc.
- *product functions*—describing the functions of a product, possibly on the side or back panels of the packages, such as the step-by-step assembly of a modular product
- *adding points of interest*—showing a beautiful flower or bouquet of flowers on a gift package, even if the product has no relation to flowers; the flowers simply create the image of a gift package
- *showing the end product*—showing a beautiful cake made out of the cake flour contained in the package, or a series of toys made from a construction kit, or the appearance of a bathroom after the product in the package has been installed there

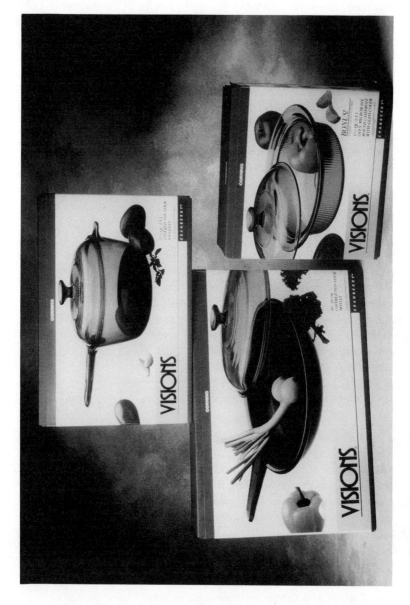

Skillful product photography can identify products inside the packages in an attractive and informative manner.

All of these applications, and many more, can be achieved through either photography or illustrations. The choice whether to use one or the other depends on a number of circumstances, ranging from subject matter, product appearance, product size and intended message by the marketer, to cost, timing and printing method.

Examples, to mention just a few, include:

- Some subject matter, such as food, is usually better portrayed through photography. A skillful, experienced food photographer can make the food in the package look mouthwateringly good. But when a single product or ingredient is to be shown on the package, a skillful illustrator can sometimes control the product visualization more effectively than could be achieved through an actual photograph of the product (see page 106).
- Small objects, such as tablets or capsules on a package containing pharmaceuticals, are usually more easily handled by illustration than photography. This is because the tiny product may have slight surface discrepancies that are not noticeable or meaningful to the product user but will be picked up by the camera, requiring expensive retouching (see page 107).
- The choice of photography versus illustration is often dictated by the printing medium. While photography can be accommodated when the printing method is offset lithography or gravure, this may not be the case when the packages are produced by flexography, dry offset or various types of screening methods. The limitations of the latter methods make photographic artwork difficult, if not impossible, to reproduce effectively.

There are many other situations in which photography or illustrations may or may not be practical, both from the point of view of visual effectiveness and due to printing and cost considerations.

These options must be weighed in advance of or during the process of designing the packages and should always be discussed with the package supplier well in advance of finalization. Neglecting to consider the medium of reproduction and its consequences on the choice of the design method can be unnecessarily costly, time consuming and frustrating to both designer and marketer.

The "Art" of Package Design

By using the elements of packaging structure and packaging graphics, the designer can draw from many sources and combine them to design a

An illustration can achieve images that go beyond photography, such as showing an orange blossom and the full grown fruit on the same branch to communicate freshness.

106

CHILDREN'S
TYLENOL®
acetaminophen
ELIXIR

Relieves children's fever and pain without

Illustrations of people are advantageous when a photograph of a person may be too specific in identifying age and personality.

package that sends the desired message to the consumer. From the package, the consumer can tell what brand, what product, what product variety, what color, what material, what size or what weight the product inside the package is. Thus, even though the consumer may not see the contents, the form and the visual appearance of the package communicates what's inside, and thus triggers or finalizes a sale. In today's marketing environment, the package *is* the product.

But package design cannot occur by itself (for design's sake or as "art"), but is closely linked to the strategic goals of a brand. Thus, the design may depend on or interact with numerous, often conflicting, elements such as company attitude or policy, target audiences, retail environments, type of product or product category, cost constraints, package manufacturing requirements, structural or visual equities, advertising and promotional directions.

This makes it extremely difficult to design an effective package because, in addition to the many external influences that guide the development of a package design, it takes a combination of many technical and visual skills, business experience, and inventive ingenuity to develop an effective package that will support the marketing objectives. In the complicated environment of today's business world, this has brought about the development of a specialist who has become an increasing influence on the development of packages: the package designer.

The Package Designer

The package designer brings to the table an unusual combination of diversified skills and experiences. These include an in-depth understanding of a wide range of marketing- and packaging- related issues, such as marketing strategy, product positioning, consumer research, graphic design, three-dimensional design, prepress preparation, printing processes, packaging materials, and many more. The package designer must keep himself up to date on a broad scope of information, ranging from trends in consumer lifestyles to the latest technical innovations and manufacturing techniques relating to packaging.

At any stage during the design development, the package designer may be called upon to apply this knowledge to a range of diversified technical and visual aspects unmatched in any other form of communication. This knowledge includes design composition, lettering and typography styles, photographic styles and techniques, illustration styles and techniques, and a variety of preparation process for printing. The designer is also expected to have technical expertise with regard to materials, material converting methods, package manufacturing methods, packaging equipment, and various printing techniques. With the giant steps taken by the electronics industry, the designer has become totally dependent on computer-generated design, typesetting, and pre-production preparation.

Because the designer's skill and experience must be so intellectually broad and technically proficient, selection of the *right* designer for any given brand identity/package design assignment is a critical link in any brand identity and packaging strategy. The sources for brand identity and packaging design are numerous and, for that reason, great differences in capability and experience among designers should be expected. Depending on company size, experience, and creative ability, these design sources can substantially affect the success or failure of a marketing strategy.

Selecting a Designer

Design consultation sources can vary greatly, ranging from a one-man/woman design studio to large consulting offices of specialists in a broad spectrum of marketing, design and technical disciplines. These disciplines may include three-dimensional design, graphic design and marketing services, brand-name development, consumer research, and a host of other services. The choice of a design consultant depends on the experience required by the marketer, the complexity of the assignment and the available budget.

Large design consulting offices usually offer a wider variety of experiences and a broader range of special skills often required by large corporations. Fees charged by large offices are sometimes, but not always, higher than those of smaller offices, and may be based on their reputation, availability of highly experienced personnel and range of services available from them.

Smaller consulting offices, or an individual consultant, may offer a more limited range of services, sometimes related to specific skills. Some offices offer graphic design services exclusively, others specialize in structural design and technical services exclusively. Few small design groups offer both, though many may claim they do. A possible advantage of a small design consulting office is that it provides a more personal service to the marketer who desires that type of relationship vis-à-vis the broader range of sophisticated services offered by larger design offices.

These are generalizations that may not apply universally, and all options offer trade-offs that must be explored and balanced in relation to the special needs of each marketer.

Advertising agencies frequently desire to be involved in package-design development, ranging from selecting and recommending a suitable design firm, to wanting to handle the brand-identity and package design within the agency by their own designers and art directors. While agencies need to be involved in the development of packaging as part of their overall strategic responsibility for their client's brand, they rarely have the internal personnel who are sufficiently knowledgeable in the particular skills and technical expertise required for package design.

Packaging suppliers occasionally have art departments of varying design capabilities. These can offer good technical expertise but, in contrast to independent design consultants and design offices, their efforts and skills are usually limited to the types of packages the supplier manufactures.

The selection process for a package designer or design firm should be based on an evaluation of their track record, their general reputation, the breadth of services they offer, their references, their facilities and personnel, and their ability to communicate, contact and follow through. The most critical criteria for selecting a designer or design firm are to evaluate the designer's ability to understand the client's marketing strategy and objectives, to analyze the retail market, and to develop unique packaging solutions that support the client's brand strategy. Based on these criteria, the designer or design firm must be able to demonstrate their ability to solve a wide variety of problems.

While it is often tempting for the marketer to seek a designer or design firm that has experience in a particular product category, it is more often to the marketer's advantage to select a firm whose overall experience lies in solving marketing-related packaging problems. This experience will act as a stimulant for creative and unique new ideas, unhampered by previous experience in a particular category.

A good designer or design firm should have such broad experience that the importance of having serviced clients in a particular product category is negligible vis-à-vis the designer's ability to understand the client's marketing objectives and develop recommendations based on the analysis of the retail market in general, and the client's product category in particular. A designer who demonstrates these capabilities should be able to develop unique design solutions in any category.

The Business of Design

It is important to realize that designers or design firms are business enterprises in their own right that understand the marketer's financial goals and responsibilities and will work with the marketer in a business-like manner. The experience and creativity that the designer contributes to the marketer's business will be provided in exchange for specified fees, just like the marketer's products are sold in exchange for a specified purchase price.

For that reason, the development of fees and related costs for a design assignment should be treated in the same business-like manner as setting the price structure for a line of products. To avoid the possibility of misunderstandings, and to protect both the buyer (the marketer) and the seller (the designer) from later disagreements, a proposal should be developed and submitted by the designer to the marketer in advance of starting any actual work. The proposal should clearly outline the procedure and methodology for the project. It should describe how the designer understands the assignment and clearly identify the work that the client should expect to receive for an agreed-upon fee.

A designer's proposal is usually divided into several incremental steps, or phases, that build on one another toward the final solution. The proposal will describe how much time each phase will take and what the fee for each phase will be. Fees are usually based on the amount of staff time required to accomplish each step of the assignment and the complexity of the assignment. Fees include such elements as time and number of staff required for design, internal and client meetings, travel, supervision as well as other activities related to accomplishing the designated task.

In addition to the fees for the designer's services, the client will be responsible for out-of-pocket expenses for purchases made by the designer on behalf of the marketer. Depending on the requirements of a given assignment, out-of-pocket expenses may include such items as travel expenses, photography, models, typography, research facilities, shipping costs, etc., connected with the assignment. These out-of-pocket expenses are charged back to the marketer in addition to the service fees. Some designers bill out-of-pocket expenses at their cost, others may add a percentage to outside purchase as a service charge, yet others invoice a combination of both methods. It is advisable for client and designer to agree on the manner in which out-of-pocket will be handled so that possible misunderstandings at the time of billing can be avoided.

Unlike ad agencies, package-design consulting firms most often work on a project-by-project basis. Some marketers prefer to "shop around" for an appropriate designer for every assignment. This is presumably done in search of design variety, but it makes an in-depth understanding of the marketer's business difficult for each new designer. The more successful client/designer relationships are likely to be based on a continuous working process.

Many designers are proud of working relationships with their clients over many years. The in-depth knowledge of the client's business, thus acquired by the designer, is a benefit to a marketer who seeks the designer's ability to relate to long-range marketing strategies vis-à-vis the marketer who sources different designers for every project. The contributions resulting from a lasting relationship with the designer could be the best investment the marketer will make on behalf of achieving long-range strategic goals and the ability of the designer to move quickly, when needed, resulting from his understanding of the client's business.

Package-Design Procedures

Just as there are many types of designers and design firms to select from, the procedures that each of them follows in developing design programs are as different as the firms themselves and should play an important part in the selection of the design firm.

There is, of course, no right or wrong procedure in developing package-design programs, but it is probably best if the procedure of the selected firm closely matches the character of the selecting client. Some companies follow product-development procedures that require careful planning of strategic objectives, attention to every small detail, extensive documentation of every step taken to develop the project, and the time required for such precise procedures. These carefully planned procedures can contribute to the feeling of security on behalf of the marketer in launching a new product or renouncing a current one. If careful procedures are characteristic of the marketer, then the designers selected to handle the package design development should have an equal passion for attention to detail.

Other companies, on the other hand, favor a highly entrepreneurial, fast-moving spirit. Such companies are less concerned with carefully documented preparation than they are with developing enthusiasm and quick, dynamic action to bring the product to market. Such a company may prefer to seek the services of a design firm of similar character, less concerned with a carefully staged procedure than with the creative excitement resulting from a "damn the torpedoes, full speed ahead" attitude.

It goes without saying that design developments resulting from such greatly divergent philosophies cannot easily be classified. For the purpose of this discussion, let us describe a careful step-by-step procedure based on the following basic methodology:

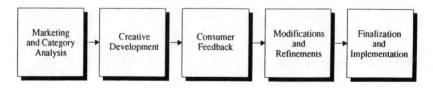

Marketing and Category Analysis

- preliminary orientation meetings with client to discuss strategy, review all pertinent factors, including marketing plans, advertising, cost parameters, etc.
- review of manufacturing and packaging facilities at the client's plant(s)
- review of existing research
- predesign research to understand consumer attitudes toward the client's product(s) and the product category in general
- category and marketing analysis relating to the brand/products
- retail audits in appropriate retail outlets, possibly in several locations
- analysis of the client's packages and competitive packages
- development of package-design criteria

Creative Development

- concept explorations
- selection of a limited number of the most recommendable concepts for further development and/or consumer feedback

Consumer Feedback

- focus group interviews or one-on-one interviews with consumers
- research analysis
- selection of concepts for further development

Modifications and Refinements

- design refinements and modifications of selected concepts
- development of packaging models or mock-ups
- postdesign consumer research to assist in selecting the final design

Finalization and Implementation

- final mock-ups or working models of the selected concept
- adaptation to additional products, varieties, sizes or package forms
- client approval process
- preproduction meetings with the client's package supplier (for

structural packaging) and/or the client's separator/printer (for graphics)
- package design finalization (working drawings and/or mechanical artwork)
- production and/or print supervision by the designer's staff
- manufacture and/or printing follow-up

It goes without saying that all of the steps in this complex procedure may not always be required or may not be desirable or affordable to the marketer. Nevertheless, in order to better understand all of the steps of this procedure, we will discuss each in greater detail.

Preliminary Orientation Meetings

Orientation meetings are the beginning of all communications between the designer and the client. They are designed to do exactly what the word says: to orient the designer regarding the objectives and parameters of the assignment and to familiarize the designer with the client's strategic objectives and whatever else is pertinent to the development of new or revised packaging. This is when the designer *listens* and asks questions before proceeding with any project development.

The more information the designer receives during the orientation meetings, the better. Not only does the designer benefit, but the effectiveness of the solutions depend, to a great extent, on the information with which the designer is infused during the orientation meetings.

The information that the designer is looking for pertains to everything regarding the product: history of the client's brand as well as competitive brands, strengths and weaknesses of the client's products, product benefits, advertising and promotional plans, cost parameters pertaining to marketing the brand and products, and any other information that may impact on marketing of the products. The designer also requires basic information about the manufacturing and packaging facilities. If the company has specific packaging machinery, this will impact the package design development and result in very precise design parameters.

Not only does the designer seek information about the product for which packages are to be developed, the designer should also be given all information that the marketer may have with regard to competitive products, competitive packaging, strengths and weaknesses of competitive products, how competitive activities affect the marketer's strategy, and where the competitive products are available.

Other information that the designer requires pertains to the anticipated target audience: age, sex and income, and any other information that may impact purchasing patterns in the category. Many times, products do not have a single target audience, but may have a primary and a secondary target audience, each of which must be considered in the design process.

The best way to communicate this information to the designer is via a "package design brief." This brief should be *in writing* and include as much detail as possible. The brief is advantageous because it provides the designer with critical information about the product and market and amounts to an agreement between the marketer and the designer as to the strategy for the design development. It has the added advantage of forcing the marketer to consider all of the marketing and strategic details that the marketer considers consequential.

Who should attend orientation meetings? It is important to understand that orientation meetings are not an idle or "political" gesture. As stated before, the information given to the designer during the orientation period impacts critically on the designer's ability to meet the marketer's strategic objectives and achieve effective package-design solutions. This means that initial orientation meetings should be attended by *all* decision-makers. If the decision-makers are not present at the initial orientation meetings, the following scenario, which every designer has experienced many times, is likely to happen: The information given to the designer did not match the strategic thinking of the decision-maker and, as a result, the design exploratory did not live up to the decision-maker's expectations. When this occurs, it usually results in major, possibly time-sensitive design changes, or even rejection of the designer's explorations. Everyone will be frustrated and no one benefits.

Another important aspect for the client to understand is that the designer must be thoroughly trusted with every detail, no matter how confidential. If the marketer feels that the designer cannot be thoroughly trusted, and, therefore holds back "sensitive" information, then the marketer should not have selected that designer to begin with. Reputable designers or design firms that have a track record of many years in business are trustworthy business partners. Without the fullest confidence of the client, the designer cannot function effectively. In any event, most major companies cover the confidentiality issue with the required signing by the designer of a legal document, the "confidentiality agreement," before any important information is imparted by the marketer.

I had first-hand experience with such secrecy several years ago, when my company was asked to design a packaging line that was presented to us by the client as a "line extension" to an existing brand. After several weeks of design development and many meetings with the client, a design concept was approved and adapted to a number of different packaging forms that made up the product line. Only after the entire program was completed and finalized did the client inform us that the product was *not* a line extension at all but a *completely new* product variety in the product category which included a new brand name. At that point, we were asked to insert the new brand name and product descriptions on the packages that had been conceived as a "line extension."

When we recovered from the shock, we felt that the real loser in this secretive procedure was the client. Had we been drawn into the client's confidence, as should have been the case, we would have approached the assignment in a totally different manner and the packages could have been substantially more effective in communicating the uniqueness of this new product. Unquestionably, the excitement of introducing a totally new product concept could have been hyped by unique packaging graphics instead of being related as a "line extension" to an existing brand.

Review of the Manufacturing and Packaging Facilities

A visit by the design staff, and/or the technical staff, to the client's facility can often be an important and educational preliminary step to a design program, particularly if the package-design assignment is of a structural nature.

A visit to the client's plant facilities will educate the designer and/or packaging technician as to the limitation as well as the opportunities of the package assignment. Even if the manufacturing and packaging facilities are not uniquely different from other similar facilities, it will assure the designer that the correct packaging procedure is being applied.

Particularly if the marketer seeks a unique structural departure from current category-typical packaging, it is essential that the designer and/or the technical design staff of the design office be thoroughly familiar with available packaging equipment at the client's plant. This will enable the designer to anticipate how much of a departure from existing packaging procedures can be conceptualized. If the proposed departures from current packaging are substantial, then the possibility of new equipment, and the associated costs of new equipment, must be evaluated. In many cases, with the knowledge acquired by a plant visit, the designer, in coordination with the packaging technicians, might be able to achieve unique design concepts that require only minor equipment modifications or minor additions to existing equipment.

At the very minimum, assuming that the package is a generic packaging structure (such as a folding carton) the designer should be thoroughly informed about the type of equipment and the capabilities of the equipment (i.e., packaging line speed, filling procedures, etc.), prior to starting any concept development. It is often helpful if a packaging engineer from the client's manufacturing staff attends the orientation meeting. This not only alerts the manufacturer's packaging personnel to possible packaging line changes, but familiarizes them with the brand strategy plan and encourages them to be available for consultation with the package designer throughout the design development process.

Review of Existing Consumer Research

If research, previously conducted among consumers of the client's products or in the product category, has been documented, the client should make this material available to the selected designer. The information contained in such research documents may or may not give the designer information that is directly applicable to the design development, but will almost invariably be informative in terms of illuminating various aspects related to the assignment.

What is most interesting to the designer is how the consumers react to the client's current packages, what visual cues they associate with the brand, what kind of product information is most important to consumers, what priority of information consumers expect on the package, where in the retail environment consumers will look for this type of product, and other information that pertains to the purchasing behavior of consumers. In the event that no such research exists in the client's file, or if the lack of product or category information inhibits an in-depth understanding of the consumer's attitudes towards the marketer's brand and the product category in general, the designer may recommend to conducting predesign consumer research to acquire such critical information.

Predesign Research

Predesign research is recommended when information about the marketer's products, the marketer's competitors, and the product category in general is not available, or if such information is incomplete or outdated.

An in-depth understanding of the consumer's behavior patterns relating to the products for which packages are to be designed or redesigned is extremely important to the designer and is used as a platform in the development of design concepts. Predesign research probes for such information as:

- What are the shopper's buying habits and claimed shopping behavior in the marketer's product category?
- What are the equities of the existing products?

- What are the strengths and/or weaknesses of the existing products?
- What motivates the consumer to buy, or not to buy, the current products?
- How does the consumer recognize and find the brand?
- What motivates some consumers to prefer competitive products?
- In what way, if any, do the packages influence the consumer's choice?
- What are the communication/information priorities that guide the consumer's buying decisions?
- What do the existing packages in the category communicate to the consumer about the products?
- What does the consumer like, or dislike, about various packaging features in the category?
- Are the existing packages easy to transport, hold, dispense, and close?
- What packaging improvement would influence the consumer's choice?

Many times when requesting such information from the marketer, the designer receives documents relating to consumer attitudes and behavior developed several years prior to the assignment. The rapidity by which the market situation changes for most retail products in today's marketing environment demands the most current state of information possible. Thus, the information that was true for marketing a brand in supermarkets only a few years ago will probably be completely outdated when the same brand is now available in the mass-market, discount outlet and wholesale market environments. The same may be true for products that are available at department or convenience stores.

If current consumer attitude and behavioral information is not available, the designer and the marketer will have to rely on past information and experience to make crucial design decisions. While past experience is a valuable component in the design development, the risk of making a wrong decision is maximized vis-à-vis decisions based on up-to-date information. For this reason, if predesign research is feasible in terms of time and the marketer's budget, the information derived from such research will be invaluable—and highly recommendable—for the marketer's and the designer's ability to precisely target the package design development.

Predesign research is most often in the form of consumer focus group interviews or one-to-one interviews, sometimes including retailers, store managers, department heads or even practitioners, in the case of products such as pharmaceuticals. This type of research, most often qualitative in nature, is meant to produce guidelines for the designer and steer him or her away from developing concepts that may be unproductive. Because of the limited number of interviews, such research should not be considered as *definite* information, but as valuable learning that often leads to solutions that otherwise may not have been considered.

Because of time and cost restraints, quantitative research is used less often for predesign learning, but is considered helpful when brand objectives or category conditions make a more extensive knowledge base desirable.

Category and Marketing Analysis

A package that successfully expresses a given marketing strategy is one where the category and marketing conditions are thoroughly understood and addressed. To achieve this, the designer analyzes the client's marketing strategy and the conditions in the product category and develops packages that take these conditions under consideration.

As discussed in the previous section, it is helpful if research in the client's files is recent enough to be valid and specifically targeted toward the product category, or if newly developed consumer research is available.

Any other documents which shed light on the product category, related product categories, and the marketer's and competitive products may contain information in which the designer may find the seeds for packaging concepts. In addition to research, this may include such things as articles in the client's file from publications that discuss the product category. It is best not to limit available information given to the designer, thereby possibly curtailing the designer's interpretation process in the concept development.

In addition to such material made available by the marketer, the designer may make a substantial effort to find any sources of information about the product category, such as information available in business publications or information obtained in libraries or from professional information services.

Equally important is the designer's familiarity with advertising or promotional plans for the brand or products for which packaging is to be developed. TV story boards, animations, concepts boards and any similar material may contain important clues leading to unique packaging concepts. Don't hold back! It is better to supply the designer with too much information than too little.

Retail Audits

One of the most important steps in advance of package design is the retail audit. The audit will take the designer to the retail environment where the client's current products are available or, if the assignment concerns the introduction of a new brand or product line, where the product is expected to be available (see page 121).

By going into the stores, the designer will become familiar with the environment in which the package will operate. This may be supermarkets, drug stores, hardware stores, retail stores, discount outlets, mass outlets, or even specialized sales environments, such as airports, convenience stores, fast-food facilities or stores that service a particular consumer, such as automotive parts stores. If the packages are for industrial products, audits are conducted at distribution centers, warehouses, trade shops, or whatever is applicable.

Whatever the environment, the designer has to be thoroughly familiar with it because the packages that are being developed must operate effectively in these environments vis-a-vis competitive products, and must take various store conditions under consideration. Shelf heights and other in-store display conditions must be investigated and stocking and restocking requirements observed.

Audits should be conducted in all environments where the products are sold, even if this means traveling to several locations. Supermarkets and other outlets frequently differ in various parts of the country. In California, for example, dump displays are a merchandising method in many categories, while on the East Coast the same products may be displayed on shelves. Also, in many product categories, regional product lines dominate certain product categories and virtually all national brands compete side by side with private labels or store brands. Many of these imitiate the leading brand's packaging in order to confuse the consumer into believing that their brands are equal to the national leader.

Designers will visit these stores, often taking photographs of the marketer's products and competitive products in the store environment. Store audits are evaluated by the design staff at the designer's office to better understand actual store conditions in the category. While in the stores, the photographer, often the designer(s) themselves, will also pick up competitive packages and evaluate their strengths and weaknesses vis-a-vis the client's packages with regard to shelf impact, brand identity, product presentation, copy, and overall design.

The retail audit is used as a continuous reference during the design-development process and is one of the most important and most informative parts of the design preliminaries. Retail audits should be made for

Retail audits are critical for familiarizing the designer with the appearance of the client's and competitive packages in the actual point-of-purchase environment.

every new design assignment. It is a mistake to rely on retail audits that were made six months or earlier because the possibility exists that the market has changed sufficiently during that period to make a potentially major difference in the approach to the design assignment.

Analysis of the Client's Packages and Competitive Packages

Another very important part in the market analysis stage is an in-depth evaluation of current packages, particularly if the redesign of well-established products is involved.

Redesign is one of the most difficult and treacherous of all design procedures. When redesigning the packages of an established product line, it must be assumed that there exists a loyal user base that marketer obviously does not wish to lose. To these users, any change in *packaging* might be perceived as a change in the *product*.

For this reason, it is critical that packaging evaluation isolate those elements on the existing packages that should be retained—most often referred to as the brand's *equity*—and separate them from those that could or should be changed.

There are numerous examples of redesign programs that have maintained a consistent look or that have changed packages almost imperceptibly over a period of many years. Examples of packages that rely heavily on long-established brand equities include well-known brands such as Del Monte, Morton Salt, Quaker Oats, Hershey Chocolate, Campbell's Soups, Wrigley, Budweiser, Tylenol, Quaker State, Arm & Hammer, Marlboro, Coca-Cola, Perrier, and even products in the otherwise highly fashion-conscious and changeable category of perfumes, such as Chanel.

But there are equally as many examples that contradict the necessity of relying on equity transfer to new or relaunched packages. A good example is the redesign of the Breyers Ice Cream packages. In this case, instead of a moderate transition from old to new design, the Breyers brand packages underwent a total design change that has little resemblance to the packages that existed previously.

In what way did the Breyers redesign differ from those that have relied heavily on equity? The main reason was that the objectives for developing new packages for the Breyers line were totally different from those that traditionally favor a slow and conservative design approach and hence retention of long-standing brand equity.

Breyers was the best known line of ice creams east of the Mississippi with a reputation for premium quality. However, Kraft, which owned Breyers at the time, felt that the geographic limitations of selling the products only in the eastern United States restricted the brand's potential. Kraft felt that the brand's premium quality would make it a good candidate for national distribution.

Retail audits, conducted by the designers, pointed to several weaknesses of existing packages in relation to the strategic goals:

- Breyers' once unique white background packages had been copied extensively over the years by many look-alike store brands (see page 124).
- Breyers was unknown west of the Mississippi. Therefore, Breyers' brand identity and packages were unencumbered by brand equity in those areas.

Based on these observations, the designers recommended risking equity in exchange for potential brand impact. This marketing strategy resulted in one of the most dramatic packaging changes in the dairy industry and perhaps in packaging in general. When relaunched, the Breyers brand name was featured in bold, contemporary lettering on a *black* background. The black background, revolutionary in the dairy industry, emphasized a larger-than-life scoop of ice cream, beautifully photographed to show off its texture and ingredients (see page 125).

After the successful introduction of the new Breyers "look" in the West and Southwest, the new black packages gradually replaced the dated-looking white packages in the East, where the equity of the existing Breyers packages was thought to be a potential problem. No such problems surfaced and the brand succeeded in its objective of becoming the leading brand of nationally sold ice cream in the United States.

This example points out the importance of retail audits and the evaluation of current and competitive packages in any given category. Observations derived from the retail audits are critical in determining the conceptualization of the packages, whether based on long-held equities or attempting new and unique directions to achieve the desired goals.

Development of Design Criteria

Following the category analysis and retail audit, and based on the marketer's marketing strategy, the designer now develops design criteria that will guide the design process throughout the brand-identity and package-design project.

Different from the strategic marketing criteria, the package design criteria outline goals that specifically address the brand-identity and package-design objectives.

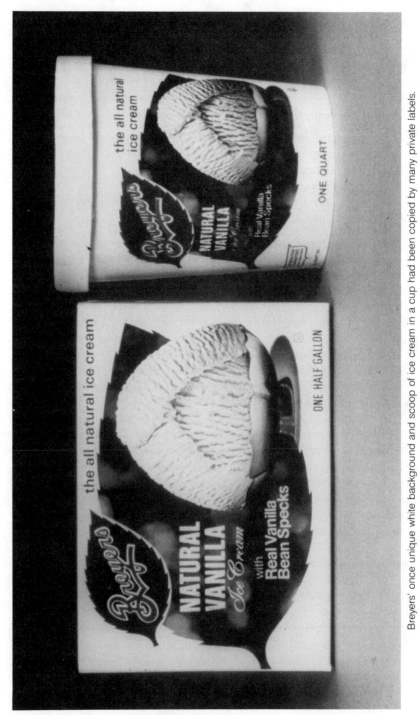

Breyers' once unique white background and scoop of ice cream in a cup had been copied by many private labels.

124

The redesigned Breyers packages emphasize brand identity and strong appetite appeal, set up against a black background, unique for this category.

These may include such considerations as:

- brand-identity requirements
- brand equity issues, if any
- product identification
- product and variety differentiation
- product visualization
- legal requirements
- structural and package form requirements

Brand-identity criteria and package-design criteria set the standard against which all package design development will be evaluated. It is important that the criteria be agreed upon by the marketer prior to the commencement of any design explorations so that the designers can navigate their design explorations within the channels determined by the design criteria.

Concept Explorations

When the marketing analysis has been completed and the design criteria approved by the marketer, the designers can start the concept explorations.

The approach to concept explorations is likely to vary substantially from designer to designer, and from design firm to design firm. Some approach the concept explorations with very precise, predetermined convictions and show the client a limited number of solutions. Others explore a wide range of design concepts. While there is no right or wrong way, a general guideline is as follows:

- If the assignment involves a simple extension to an existing line of products, such as a new flavor or product variety, it may be sufficient, depending on the criteria, to limit the explorations to a small number of concepts, such as two or three, since all of them will probably retain the existing brand identity and relate closely to the existing product line.
- On the other hand, if the assignment concerns packaging for the introduction of a new brand or product, or if marketing or category conditions require repositioning of the brand or product line, the marketer will benefit from a more extensive exploration of design concepts. It is also possible that the addition of line extensions to an existing line of products will stretch the line beyond its original intent and, therefore, require a restructuring and extensive package-design exploration.

Depending on such circumstances as the consumer's familiarity with the brand, type of product and number of products in the line, length of time elapsed since the last design, category characteristic, and marketing activities, explorations could range from design alternatives that are relatively conventional to more unique concepts that will give the marketer an opportunity to communicate a distinct point of difference from competitive packages.

Concept Selection

In most package-design programs, the original concept development is followed by meetings between the marketer and the designer for the purpose of evaluating the alternatives and selecting a limited number of solutions for refinement and modifications.

It is usually advisable to select no more than between one and three concepts for further development. If more concepts are selected, issues relating to basic brand strategy may become diffused. This may indicate a lack of a clear strategy on the part of the marketer. By selecting a limited number of concepts from a broad range of design alternatives, the decision-makers will discipline themselves to make clear decisions with regard to the strategic objectives for the brand.

When selecting design concepts, it is important that they meet the design criteria originally agreed upon while being sufficiently different to offer clear alternatives for the final package-design selection.

As an example of a meaningful selection of design directions, the redesign of the packages for the well-known line of Sudafed Cough and Cold Remedies serves as a good example. From the initial broad range of design explorations, two distinctly different package-design concepts were recommended by the designers, and agreed upon by the marketer, for consumer feedback prior to further development. Each of the two directions fulfilled equally well the desired strategy of improving shelf visibility, communicating efficaciousness, and achieving clear product differentiation. Yet the difference between the two concepts was substantial:

- A "close-in" design direction that related visually to the existing packages in consideration of transferring perceived visual equities from the existing packaging graphics.
- A concept that differed substantially from the existing packages, but more clearly communicated prescription-like efficacy.

By making a distinct difference between the selected directions, it was possible to probe and identify specific consumer preferences during consumer research. In this case, consumers expressed a clear preference for the most efficacious graphic direction. Sudafed packages have since achieved substantial success in the market and have been able to expand the line to several new products, retaining a visually cohesive line of products.

While this acceptance of a substantially new look may not be recommendable for all brand-repositioning situations, the important point to be gained from this example is that only by considering and consumer testing two distinctly *different* directions, could the final, and ultimately successful, design direction be clearly established.

Design Refinements and Modifications

Having selected a design concept most suitable for further development, the designers develop refinements and modifications leading to the final design direction. Although this phase is referred to as "refinements" and "modifications," these terms are possibly a misnomer. The fact is that modifications and refinements often require substantial additional design development to achieve the desired results. For example:

- If the product is a line extension utilizing a primary brand logo together with a secondary product logo, the potentially conflicting relationship between the primary brand and sub-brand may have to be carefully explored to find the right balance.
- If product pictures are part of the selected design concepts, a wide range of explorations may be required to select the best presentation of the product.
- A variety of products in the line may require extensive color explorations for the purpose of achieving the best possible product differentiation system.
- Copy describing the product and benefit statements that had not been available or finalized before starting design explorations may require many adjustments of design elements on the selected design direction.

Thus, brand and product identification, product description, colors, product presentation, and many other key elements may require extensive explorations during the refinement stage. It is not unusual that the refinement stage will be more lengthy and complicated than the initial concept development.

This is sometimes difficult to understand for someone who is not involved on a day-to-day basis in the design process. Perhaps it is easier to understand when one considers that, while the initial design phase was primarily concerned with generating unique design alternatives, in the refinement phase we must come to grips with specific strategic design and package production details.

Packaging Models or Mock-Ups

After the refinements and modifications of the selected design concept, it is best—even mandatory—to produce three-dimensional packaging mock-ups of the final packages, showing all structural and graphic elements. This step is necessary to make sure that the appearance of the selected packages lives up to the expectations of both the designer and the marketer and that all copy and design elements are properly positioned on the various sides of the package.

Occasionally, there is a temptation to bypass this step in an effort to save package development costs or time, especially if the package designs represent evolutionary modifications. However, packages are three-dimensional objects that should always be viewed three-dimensionally before finalization. For example, the visible surface of a can or a bottle label may represent only 40% of the total display area. The remaining 60% recedes around the sides and disappears on the back of the circular package. While the initial design explorations can be done in the form of *flat* presentations of the primary display panel, it is virtually impossible to accurately visualize the entire packaging unit until the label has been wrapped around the can or bottle, or all display panels of a carton can be viewed.

Thus, the three-dimensional mock-ups ensure that important copy and other design elements appear within the 40% visible label surface and that all other elements are properly positioned on all other sides of the package. Potential errors in judgment, such as brand name or product description wrapping around the circular label surface too far to be seen at one glance or, copy placed incorrectly on the designated carton panel, can be discovered and corrected at this point.

Not only should round cans be mocked up, but rectangular cartons, blister packs, pouches, and bags should all be reviewed three-dimensionally. A three-dimensional carton with top and side showing may give a substantially different impression than viewing the primary display panel alone. For blister packs, the overlap of the blister flanges and the products contained in the blisters may have to be considered in regard to the readability of copy and other visual material. Also, the appearance of flexible bags and pouches can look substantially different when filled with actual product. Large paper bags, such as those used for pet foods or sugar, are often displayed horizontally so that the end panels become more important than the primary panels under actual display conditions. These are but a few examples of why all package designs should be reviewed carefully in the form of mock-ups to make sure that they function properly at the point of sale.

Needless to say, any three-dimensional structures, such as bottles or jars, must be mocked-up three-dimensionally to be reviewed and evaluated properly. Using wood, acrylics, plastics or any other material, these three-dimensional mock-ups can be developed either by the packaging designer in coordination with the supplier or by a specialist model maker. Model making requires sculptural skills, materials knowledge and special equipment. Some designers have their own model shops.

Mock-ups are also frequently required for the purpose of consumer research to assist in the selection of the final package. In any event, the importance of developing three-dimensional mock-ups to visualize the final appearance of a package or a line of packages, whether graphic or structural, cannot be overemphasized. The time and costs for developing three-dimensional packaging mock-ups will help minimize or avoid more costly and time-consuming errors that would otherwise almost certainly occur.

Package Design and the Computer

In the last few years, the use of the computer in package design has almost totally taken over from the previous method of developing packaging. Package designers use computer technology for virtually every aspect of their trade, ranging from concept development, to rendering, to mechanical artwork and every step in between—including scanning in of photographs, developing illustrations, setting type, creating logos, retouching, and simulating mass displays in the store environment.

The role of computers in package design has been both a blessing and a curse. The blessing is the designer's increased ability to develop and render a wide range of designs without having to allow for lengthy rendering time of the concepts. Experimentation with designs, colors, typography and other design elements has been simplified. A plethora of type styles are available on computer programs that can be molded, curved, stretched, condensed and made heavier or thinner at the designer's will. Color and copy changes can be handled with comparative ease and, in almost a moment, compared to earlier when copy had to be ordered from typographers and colors had to be painstakingly hand-rendered—all taking from one to several days (see page 131).

Computer-generated design, from logo design to pre-press mechanical artwork, has become the universally accepted methodology for package design.

Most designers are pleased with designing on computers, as it gives them the opportunity to use their time for greater experimentation, rather than having to spend hours upon hours just rendering their concepts for presentation to the marketer.

But computers are also a proverbial curse in that they establish a false sense of security and fee structures. Marketers expect faster turnarounds of design development and design changes because they visualize that this can be accomplished in minutes. While this may be true in some cases, it forces an anxious designer to make shortcuts that may not necessarily benefit the design development and, therefore, may ultimately not be to the marketer's advantage.

Among marketers, there is also the impression that the computer makes designing easier, even routine. Unfortunately, this notion is encouraged by unscrupulous, or untalented, design sources who promote themselves as being able to produce designs quickly and cheaply through the use of computers. Nothing is further from the truth. The computer is an electronic tool that only produces what the designer puts in. The computer output of a talented and experienced designer will be unique and effective. A bad designer's computer will issue only poorly thought-out and poorly executed designs, negating the speed and lower fees resulting from such a procedure.

There is also a false impression among some marketers that the speed of the computer in designing and making design changes should result in cheaper production and design costs. This thought process does not recognize the substantial investment that the designer is obligated to make in the basic hardware and constant upgrading of such hardware, as well as the high cost of acquiring computer programs that must be constantly upgraded by better, more flexible and faster new programs. The designer's investment in computer hardware and programs is substantial and continuous and must, logically, figure into the designer's overhead costs which are part of their trade and fee basis.

Consumer Research

Consumer research is often an important step in the development of packaging programs. Brand management of many companies, as well as design consultants, utilize consumer research to develop design strategy and design directions, to confirm the appropriateness of the selected designs in regard to the marketing strategy, and to discover any potential negatives that may have escaped their attention during the brand-identity and package development.

Not too many years ago, marketing management did not utilize consumer research extensively in connection with package-design development because they were either not sufficiently familiar with package-related consumer research techniques or they felt that their decisions, based on their intuitiveness and marketing experience, were sufficient to select the appropriate package-design directions.

Package designers, in the past, were even less likely to utilize or recommend consumer research for their packaging concepts. In fact, many package designers were vehemently opposed to any type of research, which they felt would inhibit their creativity, leading to conventional package design solutions. To this day, many package designers still voice doubts about the value of packaging research and defend their experience and creative capabilities as the preferred method of selecting package designs. These designers feel that subjecting their creative recommendations to mechanical eye tests and consumer interviews by researchers, who may be more oriented toward statistics than creativity, could channel the design solutions into parameters that favor conventionality over invention and uniqueness.

These concerns are not necessarily totally unfounded. Many research firms that may be thoroughly experienced in consumer research of advertising are not necessarily sufficiently acquainted with probing for the peculiarities and characteristics of *brand-identity and packaging* communications. The difference is primarily that advertising, especially TV commercials, has the ability to utilize many tools—such as music, facial expressions, voice-over, and other visual and audio effects—that are not available to packages. Also, commercials and print ads, in contrast to packaging, need not compete for attention side by side with competitive products, and they are not subjected to the need for long-term performance.

In contrast to the excitement that can be created in a TV commercial, packages are static elements that have to achieve an entirely different set of objectives:

- Packages sit on shelves, hang on peg-board walls or are otherwise displayed at the point of sale.
- Packages usually are next to one or more directly competitive products and must be distinguishable in the competitive array.
- Packages must communicate their contents clearly and quickly, usually within a very restricted label area dictated by the package proportions.
- Packages must create memorability in the consumer's mind through emphasis on brand identification and their overall visual appearance so that the consumer will find them easily and repurchase the product.
- Packages have to deal with an entirely different array of technical requirements, relating to production, printing, assembly, protection of the products inside, adaptability to the manufacturer's packaging equipment, shipping and warehousing, to name just a few.

A few research firms, such as Perception Research Services and Opatow Associates, that specialize in package design research and The Consumer Network, which conducts focus group and package satisfaction studies, have developed research techniques geared to probing the appropriateness of the packages in relation to the marketing strategy, assisting marketers and designers in the decision-making process by providing guidance for modifications and further refinements, and to strengthen the communication effectiveness of the packages.

Despite previous and even current objections to consumer research by some within the design community, many design consultancies have learned to appreciate the need for an unbiased, consumer driven opinion research of their creative output. A few have actually added research capabilities to their own services.

The research involvement of design firms fall roughly into three categories:

- Design firms that rely entirely on and recommend to their clients outside research specialists who develop research methodology, conduct the research procedures, and develop the research reports.
- Design firms that have a research director who is responsible for selecting outside research services from a wide array of possible research techniques, coordinating with the client and the research firm the development of research methodologies, observing the research procedures, interpreting their final reports, and making summary recommendations relating to the marketer's strategic objectives.
- Design firms, such as Gerstman + Meyers Inc, that have set up their own, semi-independent research division to handle consumer feedback in close coordination with the design firm's design assignments. In these cases, the primary objective of internal research capabilities is to be able to evaluate design assignments and to guide the staff in appropriate design directions even before presenting their creative output to their clients.

The selection of appropriate research firms and research methodologies for package design research depends on the type of feedback desired to make design-related judgments that are not subjected to personal prejudices by individuals involved in the design-development process, both on the marketer's and the designer's side.

As covered in Lorna Opatow's chapter, "Getting It Right," many different research methodologies may be used for brand-identity and package-design research.

Tachistoscope (often referred to as T-scope)

A methodology where individual packages, or packages shown in a competitive array, are exposed to the consumer at various short time intervals to test for shelf impact, brand and product recognition and findability of various package-design alternatives.

Eye-Tracking

A method utilizing laser technology to trace the path of the eye as it surveys a package or a shelf display. This method provides a diagram that shows the movement of the consumer's eyes as they moved from design element to design element, or from copy element to copy element on individual packages or from package to package in the competitive array (see top of page 136).

Focus Group Interviews

Focus group interviews, as the name implies, involves interviewing small groups of consumers—usually a representation of product users and nonusers in a specific category—by a moderator (such as editor Mona Doyle), skilled in eliciting from the consumers, opinions and attitude toward the category in general and various packages in particular. This method is popular among marketers and designers because it allows for a great deal of flexibility and cost effectiveness, and because it can be observed and listened to by the marketer and designer in an adjoining room behind a one-way mirror (see bottom of page 136).

One-to-One Interviews

Similar to focus group interviews, one-to-one interviews attempt to solicit from consumers opinions and attitudes toward products, product categories or package-design directions by individual interviews of appropriate respondents (i.e., users and/or nonusers of the products). These interviews can be conducted in or near the shopping environment or at separate premises, depending on the type of products involved. One-to-one interviews are particularly applicable when the respondents are individuals who have limited time for participating in such procedures—such as doctors—or individuals whose relationship to certain products touch on intimate or sensitive issues—such as, for example, drugs for AIDS patients.

Simulated Store Tests

This method requires simulation of an entire store section or product category with the package-design candidates displayed among competitive or potentially competitive products. Respondents are asked to "shop" the section, and their shopping behavior is noted by the observing researcher. This method includes interviews of the "shoppers" who are

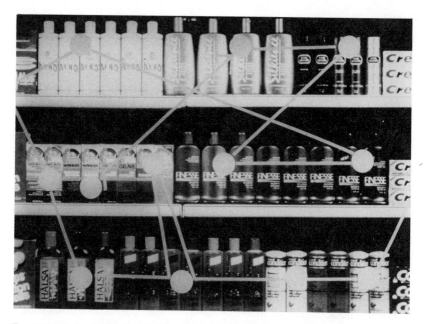

Eye-tracking documents the "seeing experience" (which items are seen and which are missed). (Photo courtesy of Perception Research Services, Inc.)

Focus group interviews, led by a skillful moderator, probe consumer opinions and attitudes to steer package design to achieve the desired strategic objectives.

asked to comment on what they have selected (or not selected) and explain why they made certain choices with particular attention, of course, to the proposed design concept.

Test Markets

The most realistic, and by far the most expensive of research methods, is utilization of actual test markets. To achieve realistic shopping situations in the test stores demand (a) actual production of a large number of packages, filled with actual products, (b) careful selection by the marketer of the most appropriate test environment, and (c) agreement of one or more of the marketer's customers to use their stores as testing facilities. While this testing method is the most realistic, it requires a massive effort on the part of the marketer to organize. As a result, it is also the most expensive testing method and usually requires a lengthy period of time—often a year or two—to observe and monitor the movement of the test packages.

When a package redesign is consumer-tested, the current packages should always be included in the testing as a benchmark against which the respondents' reactions to the new packages can be compared. However, it is generally not recommendable to ask respondents to evaluate the current package designs in direct comparison to new designs, as this would never occur under actual shopping conditions.

Respondents should never be encouraged to approve or disapprove package designs directly. Questions such as "How do you like this color?" or "Which design do you prefer?" should never be asked since this does not simulate real purchase situations. Answers to such direct questioning will be misleading, making the respondents, in effect, design directors—a qualification that is inappropriate and unlikely in real purchasing decisions. Instead, questions should always be directed toward the *product* so that the respondent's comments will indicate whether the package design relates, or does not relate, to the intended marketing strategy.

The effectiveness of directing attention to the *products,* rather than the packages, can be illustrated by the following example:

Several years ago, Narragansett Breweries, a regional brand sold in the Rhode Island/Massachusetts area, consumer-tested design concepts by producing actual printed cans for each of three design concepts. All cans contained the same beer. The testing took place among beer drinkers in the regional area where the beer was well known. Respondents were asked to taste-test the beers in the three cans and describe their opinions about them to the researchers. The opinions expressed varied substantially, ranging from "light" to "heavy" and from "watery" to "full bodied" (see page 138).

Taste preference was achieved by "taste testing" (actual design testing) three different packages, all containing the same beer. The center design was identified by beer drinkers as a stronger, less watery, higher quality than the others.

Since the beer in the three cans was, of course, exactly the same, the responses showed conclusively that the packaging graphics influenced consumers' perceptions of the products. This was exactly the information the marketer was looking for. The winning design was eventually produced and marketed, based on the comments expressed during the research.

Consumer research has become an important segment in the package-development process. While it does not guarantee conclusive decisions, it can be extremely helpful in focusing on design objectives and their appropriateness for a specific marketing strategy. While consumer research does not necessarily determine the final design direction, consumer research can, and usually will, lead to specific recommendations.

At the very least, packaging research will confirm the appropriateness or inappropriateness of the verbal and visual communication elements of the designs. For example:

- For a long time, black was a no-no for packages containing food. This was especially true for dairy products, which were usually associated with white because of their milk content. Thus, the black backgrounds of the Breyers ice cream packages, was initially considered to be potentially risky by Breyers' brand management. However, packaging research conducted independently confirmed the designer's recommendations and the Breyers brand subsequently became the leading nationally sold premium ice cream in the United States.
- Pet food and baby products packaging frequently feature pictures of animals and babies that are meant to appeal to the empathy of the purchaser. These pictures, therefore, are critical to the potential sale of the products. Since likes and dislikes of pets and babies are subjective, package design concepts for these products are frequently researched to obtain consumer feedback that is not tainted by the marketer's—or the designer's—personal preferences.

The same procedure used in packaging research, when applied to structural packaging development, often leads to refinements and modifications. The shape of a bottle, the placement of a handle, the ease of opening and closing, the size perception and the tactile feel of the packaging surface or shape all contribute to the acceptability, or nonacceptability, among consumers.

One aspect of research that has not been discussed is the ever-present possibility that package designs being tested will be totally rejected by the respondents as being unsuitable for the products. While this is not a pleasing thought to either marketer or designer, and while it does not occur frequently, there is always a possibility that the judgment of both the marketer *and* the designer has been channeled in the wrong direction. No matter how difficult it is for the marketing management and the designer to accept such a reversal, it is obviously better to find this out at the consumer research stage than to market products in packages that may subsequently fail.

Thus, consumer research can be a "disaster check," which can prevent the marketer from introducing products in packages based entirely on intuitive decisions that risk failure. Most importantly, since total failure of a design program often indicates problems beneath the surface of the marketing strategy, consumer research can be helpful in reevaluating marketing strategy and steering the package design development in directions that will be a better fit for the marketer's brand and products.

In any event, consumer research never imposes final decisions upon the marketer, who always uses his/her own judgment to determine the direction the brand will take. While consumer research is a valuable tool, the ultimate choice of which direction a brand and its packaging will follow should always be based on the marketer's vision of long-range marketing strategy.

Design Finalization

The mock-ups or working models utilized for consumer testing were very likely prototypes that will require refinements based on the interpretation of the packaging research. For that reason, and for reasons discussed earlier, it is recommended that final mock-ups or working models be prepared so that the marketer can visualize the appearance of the really *final* packages suggested during the consumer research. These final mock-ups or working models should be as real as possible, including all copy and design elements.

In the event that the product line consists of a variety of sizes, flavors, or different types of packaging forms, it is recommended that final mock-ups of all of these be reviewed in order to accurately be able to evaluate the appearance of the entire product line prior to production. These mock-ups can also serve a number of other purposes, such as circulating them among various offices for their approvals, as well as serving as temporary stand-ins in the preparation of advertising and sales promotion material at a time when actual packages may not yet be available.

Client Approval

In many cases, especially in large corporations, the approval process can be a complicated and time-consuming one involving marketing, legal, purchasing, manufacturing and other departments, all of whom must approve and sign off on the final designs with regard to their respective special concerns.

During the client approval process, it is important that packages be viewed within the competitive array. Thus, approving packaging on a conference table is not a recommended method for visualizing final packages. Either the final mock-ups should be taken to a store, or a simulated shelf display should be set up at the marketer's office. At the very least, the selected package designs should be reviewed among other packages with which they will compete at the point of sale.

Once the approval process has been completed, the designers are in a position to prepare mechanical artwork for the packages so that they can be produced in accordance with the appropriate manufacturing or printing process.

Package Design Finalization

Finalization of packages is as important a step as the design itself. Finalization is the technical portion of the design program. The skill of finalizing the structural details or the printed elements of the package is critical to achieving excellence for the final package appearance.

There is a substantial difference between the design development of the package and the finalization: During the design stage, it is easy to shift gears, make changes, refine various details of the design, reduce and enlarge design elements, adjust copy elements, modify colors—or even rearrange entire portions of the design.

Finalization of packages does not have that luxury. If package development is structural in nature, the working drawings and blueprints for the final packages are critical. The slightest error in dimensions may affect the content of the package, the structural rigidity of the package, the weight of the package, the handling of the package, and the cost of the package. Material specifications have to be correct to hold, store, and protect the product. For example, bottle and jar finishes have to be precise to fit the caps or other types of seals and closures to avoid any kind of leakage. Package proportions have to be precise with regard to stability, content, and label surfaces (see page 142).

The same attention is required for print production. The preparation for print production involves many steps, including exact specification of the print area, color bleed specifications, type specifications, package copy, proofreading, rendering of logos, preparation of photography and/or illustrations, color specifications, and many other similar details.

When photography is part of the design, this alone is a complicated and sometimes lengthy process that requires specific expertise. Preparation for

Precise specifications in structural package design development is critical to production and marketability of the product.

photography may include searching, finding and selecting props, such as dishes, utensils, backgrounds and special effects. If people, or any parts of people are shown in the photographs, the designer must interview models and select those most appropriate for the subject matter, as well as selecting hands, faces, hair or other human features that may be essential to the package design. Last, but not least, contractual arrangements must be concluded with the models or model agencies, and contingencies or buyouts must be agreed upon to allow the continuous use of the photographs on the packages.

Designers must interview photographers, food stylists, fashion stylists, and room stylists. If the products are of a technical nature, interaction with appropriate technical personnel may be required. There will be discussions about how to show the product, how to light it, how to emphasize or deemphasize details.

Following photography, there is the lengthy process of selecting the best transparencies and retouching these, or making arrangements with the platemaker to handle the retouching through computer-generated methods. In some cases, several photographic components may have to be combined. Each step involves the combination of marketer, designer, and platemaker to achieve the desired effects.

The same painstaking process applies to the final preparation of hand-rendered, or computer-generated illustrations. Illustrations are most often included on packages when certain visual effects are desired that cannot be achieved with photography, when important product features cannot be satisfactorily shown by means of photography, or when product use or preparation instructions need to be visually communicated. In many instances, the decision to use photography or illustration is simply the designer's choice of style.

When all of these details have been completed, when the artwork has been finalized, proofread, approved by the client and the disc containing the final artwork forwarded to the platemaker or the supplier, there is no turning back. Finalization of packaging, whether structural or graphic, means: The buck stops here.

Preproduction Coordination

An important step that is often neglected, or acted upon too late by many marketers, is a preproduction coordination with platemakers and packaging suppliers. Preproduction coordination involves meetings in which the people responsible for the *production* of the packages, the *purchasing* of the packages and the *filling* of the packages on the production lines meet with those responsible for marketing and package design. These meetings are critical because discussing the recommended designs prior to production can prevent possible complications and costly delays during the package production process.

In most design-development programs it is advisable to arrange for pre-production meetings as early as possible. Designers are responsible for developing unique, new package design concepts in order to achieve a distinctive marketing position for the marketer's product. To achieve this, their design development may include concepts that do not completely adhere to the existing production capabilities and procedures at a client's plant. For this reason, it is important to discuss production parameters during the early stages of design development, or even *before* the start of the design development, in order to be familiar with the marketer's existing packaging equipment, budget requirements and physical requirements for the packages. These may relate to such important issues as the preservation and protection of the products or the interaction of packaging materials with the products inside. This especially applies to the packaging of foods, liquids, pharmaceuticals, chemicals, fragile products, products sensitive to light or atmospheric conditions, etc. Also, unfortunately, the recent phenomenon of pilferage at the point of sale often affects the design of the packages.

If a new or modified packaging structure is involved, the manufacturer who produces the packages—whether these are folding cartons, polyethylene bags, plastic trays or glass bottles—should have an opportunity to interact with the designer. No matter how much experience the designer may have with materials and production of packages, early discussion frequently leads to suggestions by the supplier that will improve the final production results, including the possibility that the designer may have overlooked, or was unaware of, important production-related details. Pre-production meetings can often simplify the production of the packages, resulting in time and cost savings, avoiding misunderstandings or misinterpretations on the part of any participant in the package design development process.

On the other hand, there is sometimes a tendency among the production staff of the product manufacturer to suggest design changes which, while making production procedures easier for them, may compromise the intent and character of the design or make it unnecessarily difficult to achieve a unique design solution. In that event, the designer should have an opportunity to convince the client's technicians and the supplier to produce the packages in the manner that will achieve the recommended design. Or, the designer may be able to achieve the desired uniqueness through minor design modifications that will simplify the package production without seriously weakening the design concept.

When print production is involved, it is equally important for the printing supplier and color separator to meet with the designer well ahead of package finalization to review the designs and discuss any possible print-preparation or print-production problems. The number of colors, for example, relates directly to the equipment available at the printing plant and the printer's capabilities in reproducing the required design elements. Each printer has different printing equipment and different printing procedures.

The importance of being familiar with these at the early stages of the design process cannot be overemphasized. This is especially true in today's print preparation procedures. Not too many years ago, all print preparation was done from reflective artwork—mechanical artwork prepared in the form of black and white paste-ups of all copy and all design elements on flat boards. These were photographed by the platemaker and printing plates were prepared through various photochemical processes. Today, with fewer and fewer exceptions, all mechanical artwork is prepared by the designer or the designer's staff on computer discs, which are turned over to the platemaker and directly integrated into the plate-preparation process.

In many cases, platemakers have their own staff to prepare computer-generated mechanical artwork based on design concepts prepared by the designer, with the designer taking the role of a consultant rather than a producer of mechanical artwork. This, once again, underlines the importance of preproduction coordination with the producer of printing plates and the printing supplier.

In addition, many packages marketed in very large quantities may be produced by more than one printer and in different locations so that the mechanical artwork has to satisfy the preprint facilities and printing procedures of several different suppliers. The same holds true when product lines include a variety of different package forms. For example, a large line of household gadgets may include a virtual arsenal of package forms, including blister cards, plastic bags, folding cartons, clam-packs and many others. Similarly, a line of beverages may be available in aluminum cans with dry-offset printing, plastic bottles with paper labels printed flexographically, glass bottles with screen printing and paperboard multipacks printed by the lithography process. All of these require different print preparation methods, and all of them must result in a cohesive visual program.

The people responsible for purchasing packaging and those responsible for running the packaging lines at the marketer's plant should also participate in preproduction meetings. They too, have critical responsibilities in making sure that the print production of the packages will be handled efficiently and economically. Their inclusion at an early stage of the design development may avoid serious production complications later on.

Thus, prepress meetings are critical to package design finalization. Although it is sometimes difficult to get all the parties together at one location, prepress meetings should never be neglected, or even bypassed, for reasons of expediency. The time and money saved in ignoring prepress coordination may lead to costly mistakes and, in extreme cases, jeopardize the launching of a product at the scheduled time.

Production and/or Printing Follow-Up

When we have finally arrived at the point at which the packages are being produced or the labels printed, it must be recognized that this is, again, a technically precise procedure. Yet, much in this process also depends on personal judgment. It should never be taken for granted that the packages, no matter how carefully and expertly prepared, automatically come out exactly right. Careful supervision during the manufacturing and printing process is as critical as every other step in the design and finalization procedures.

If the marketer of the products is a large corporation, it is likely that their personnel includes specialists who are responsible for supervising packaging manufacture. These specialists will probably be at the manufacturing plants when the packages are produced or printed and will be responsible for approving the final packages. This procedure may take several days because of the possible need for last-minute manufacturing adjustments during the production run, or even occasional breakdowns of the supplier's machinery.

Designers are frequently required to participate in the approval process and they are usually eager to do so. Participating in the production supervision gives the designer another opportunity to interact with the final package development. An experienced design firm includes technical staff who understand packaging manufacture and packaging printing and "speak the same language" as the supplier's technical personnel. They will work hand in hand with the packaging manufacturer to make adjustments in the structural or printing details. When this occasionally requires trade-offs, such as the balancing of colors that may be difficult to reproduce, the designer may be in the best position to visualize the results and make productive suggestions to solve the problem.

Thus, even this very last step in the package design process—the manufacture or printing of the packages—is an integral part of meeting the marketer's brand strategy. From the very first meeting to the final production of the package, the designer's involvement is crucial to achieving the best possible results.

The Role of Package Design

The role of package design is to support the strategic objectives for the sale of the marketer's products. What we have just discussed is the interaction of various steps in the package-design development process.

The constant changes in market situations and marketing procedures and the rising costs of advertising have catapulted packaging into a primary position among strategic support media and, in a few cases, has even reversed packaging's previous subsidiary role, vis-a-vis advertising, to a leadership role. In many examples, packaging serves as the sole medium for sales successes.

The changing market has had a substantial influence on the way packaging is utilized as part of the marketing strategy. Package design today plays a crucial role in focusing on the specific position of the product in the product category. With product positioning becoming increasingly more targeted to appeal to specific consumer segments, packaging has to play an ever-more critical role in precisely communicating to the target audiences.

This is most noticeable in the way packaging appeals to a growing number of diverse consumer groups, such as the growing market of working singles who have more discretionary income but less time to spend it, and the greater worldliness of Americans, many of whom have an opportunity to travel extensively and have developed more sophisticated tastes and habits.

Even the substantial changes in our social fiber will influence package design, including the growing population of elderly people, and greater concern for health-related aspects, such as the chemicals, fat and nutrients utilized in our food and cosmetic products. The growing chasm between rich and poor in the United States has also contributed to an ever-greater scope of products and packages, ranging from luxury brands to economical store brands for everything, including food, health care products, cosmetics and even such items as housewares, hardware, cameras and electronic equipment.

An equally important influence on packaging is the availability of technologically oriented products, such as the ownership of microwave ovens by larger segments of the population, the increasing number of personal computers and related equipment, as well as a powerful do-it-yourself market. Products previously sold through mail order are now being offered in retail stores, requiring their packaging to change to enable these products to sell themselves on the store shelves. Also, an ever-increasing number of pharmaceutical products, previously available only through prescription, are now available as over-the-counter products, requiring appropriate consumer packages.

Home shopping programs, currently geared mainly to household products and luxury items, have already found a large and enthusiastic buying audience, and the time when television-transmitted sales of food, pharmaceuticals, household goods, hardware, etc., is probably soon upon us and will undoubtedly change the appearance of packages dramatically.

Because of the diversity of the markets and the growing need for precise positioning and product strategy, it is important to realize that package design requires an equally diversified approach. Design procedures may vary considerably based on objectives that go beyond redesigning packages for an existing product line, particularly products that are well known and have developed a strong franchise in the consumer's mind.

When designing packages for line extensions, critical decisions have to be made as to whether the graphics should be closely allied to the existing packages or whether the line-extension products are sufficiently different to make a unique statement of their own, and thus may need to identify these differences through a different design strategy. Thus, changes in product positioning for the purpose of reaching a different target audience require the marketer and the designer to explore packages that range from retaining current equities to designs that communicate the unique characteristics of the product to the desired, new target audience (see page 149).

The role of package design is critical in communicating marketing strategy. Faced with the fact the supermarkets in the United States today carry from 20,000 to 30,000 products—and mass merchandisers even more—brand and product managers must recognize that, in order to succeed in this tough competitive environment, they need to determine precise targeting objectives for their products, which will in turn become the foundation for the packaging-design development. Only then will the package designer be able to effectively assist the marketer by developing package-design concepts that are precisely targeted and support the strategic goals of the marketer.

In their book *Positioning,* Al Ries and Jack Trout state: "Every drug store and supermarket is filled with shelf after shelf of half successful brands." No marketer will consider "half successful" a satisfactory situation. When it does occur, chances are that one of the reasons for the limited success is that the marketing strategy is not precise. Based on a weak strategy, packages cannot effectively communicate, no matter how cleverly designed.

Packaging is able to support the marketing strategy by focusing on consumer targets ranging from a broad, general audience to specific target groups. There is no need for products to be "half successful." Supporting

Breathing new life into a static category, Schick's new brand identity and package design creates the excitement of a high-tech product.

marketing strategy with the appropriate packaging strategy is possible if
we remember the following rules for package design development:

1. The key to creating packages that effectively support a marketing strat-
    egy is an in-depth understanding of the current market and consumer
    attitudes in the product category. To accomplish this, a thorough
    marketing analysis, including retail audits, packaging evaluation, ex-
    ploratory research and a precisely targeted design criteria, must
    precede each design development.
2. To develop unique and effective packaging concepts, a wide range of
    design options, within cost, manufacturing, and marketing parameters,
    should always be explored. The most successful brands have been
    those that had precisely targeted strategic objectives and were not
    timid about communicating these to the target audience.
3. In redesigning packages of established products, a clear understanding
    of the relative importance of visual equities must be established. If nec-
    essary, consumer research should probe for such issues so that an in-
    telligent, unemotional and unprejudiced determination can be made
    as to how to communicate the intended marketing strategy most effec-
    tively.
4. To achieve the best results, the package designer should be regarded as
    a trusted and integral associate of the marketer, not as an outsider to
    whom information is parceled out selectively for confidentiality
    reasons. Just like other professionals such as lawyers, accountants or
    doctors, professional designers respect their client's needs for confi-
    dentiality and realize that any breach of that confidentiality will cause
    substantial damage, not only to the client, but to their own profes-
    sional reputation.
5. The same time and care that is applied to development of the products
    themselves should be applied to the development of packaging solu-
    tions, whether structural or graphic development. No matter how
    unique or superior the product may be, only through the fullest atten-
    tion to the package development, from conceptualization to finaliza-
    tion, can the package function effectively to give the product the sup-
    port it deserves.

The role of package design in relation to marketing strategy is crucial in
today's market environment. The design of the package brings together, at
a single glance, all the factors of the marketing strategy. The ability to communicate the marketing strategy to the consumer is the ultimate role of
package design.

HERBERT M. MEYERS, FPDC

Herbert M. Meyers has been a specialist in branding, marketing, corporate identification and package design for major corporations and design firms since 1955. He joined Richard Gerstman in 1970 to form Gerstman + Meyers Inc, now a leading international consulting firm, specializing in brand identity and design.

The firm assists corporate, industrial and consumer product firms in understanding and shaping the identities of their brands through the development of proprietary strategies and systems. Over 100 major corporate clients are served in the U.S. and internationally, with affiliates in Switzerland, Canada and Brazil.

Mr. Meyers is a frequent lecturer and writer on branding and packaging subjects and is the recipient of numerous design awards. He is past president of Package Design Council International, a member of several business and design organizations, including PDC, AIGA, APDF, AMA and PDA-Europe, and also serves on the Board of Trustees at Pratt Institute.

The Consumer Side
of Packaging Power

A strong package is powerful today and is likely to be even more so in
the more person-responsive marketplace of the twenty-first century. Consumers are increasingly adept at making purchase decisions based on information available on the package or at the point of sale. They are making
more comprehensive marketplace decisions about what works best for
them and what is best suited to their needs. In keeping with this behavior,
they rely on their own packaging needs or preferences. Thus, they
routinely make lifestyle and use-benefit choices between canisters and
boxes of raisins, plastic-handle jugs versus glass bottles versus gable-top
paperboard containers for milk and juice, and canned versus jarred
sauces. They read and act on package descriptions of package and product
attributes. At check out, they choose between paper and plastic grocery
bags, or ask for both.

When a new package communicates a meaningful improvement to consumers, they are quick to make the switch. An appealing new package can
cripple the pull of the "old" packages and trigger a competitive stampede,
resulting in a whole aisle of new products and repackaged formulations in
the package of choice. Examples of these packaging dominos are plentiful.

- As soon as consumers showed that they were willing to carry
  12-pack "cubes" of softdrinks to save time and money, all the
  major players introduced 12-pack and 24-pack cubes and Pepsi
  introduced a 30-can multipack called Block Party. Within a one-
  year period, cubes and large multipacks became a regular
  feature of the beverage aisles in supermarkets and mass
  merchandise and convenience stores.

- As soon as consumers showed that they had a decided preference for plastic bottles of mouthwash (and almost every other product designed for bathroom use), everyone (but Listerine) played follow the leader and the category switched from glass to plastic, leaving Listerine to its unique glass-in-corrugated packages until more and more consumers showed that they were disenchanted with multilayer packaging, and Listerine capitulated (see page 157).
- As soon as susceptor technology made microwavable pizza and popcorn possible, dozens of new susceptor-packaged products were created, all of which were more convenient than delicious.
- As soon as aseptic drink boxes started selling juice and juice drinks to parents of young children, every ready-to-drink children's beverage introduced a juice box product. A whole new sub-category was created.
- As soon as consumers showed a strong preference for one-handed flip closures on hair care products, most national brands adopted them and most store brands followed. Flip-top closures quickly became the standard closures on all but single-use trial-size hair care packages.
- Very soon after toothpaste pumps attracted a large number of consumers who disliked the mess or the process of squeezing, a proliferation of pumps gave way to a proliferation of other alternatives, like flip-caps and stand-up tubes.
- Almost as soon as minor player Minnetonka proved that a lot of consumers wanted to buy hand soap in a pump dispenser, all of the major soap producers brought "new" liquid soap products to the market.
- As soon as possible after Sargento showed that consumers moved in droves to their easy-to-open, easy-to-reseal packages of shredded cheese, "new" shredded cheese products in see-through resealable packages began to appear. In a very short time, the dairy case had a whole new look, filled with dozens of new products. Most were pegboarded and provided quick and tasty ways to help hurried consumers assemble a fast meal or add a creative, tasty touch to tacos, pizza, salads, nachos, or "speed scratch" recipes.

  Shredded cheeses have actually been around for a long time—Sargento introduced a shred for tacos in 1971—but consumers didn't get excited about it until the convenience of shredded cheese was showcased by an easy-to-open-and-reclose see-through package. Consumers who discovered it literally reacted with glee. One asked: "Why isn't everything packaged this way?" Cheese packagers got the message well enough to start introducing new products in the packages that consumers purchased, but the dairy industry still credits the growth of "shreds" to consumers' "growing interest in ethnic cooking."

From our research with consumers, we know that the easy-open, resealable "shred" package played a central, rather than a peripheral, role in the growth of the shredded cheese category.

A package that solves a problem is a package that's good enough to talk about.

A package that's talked about generates a lot of product trial—enough to create a whole new category, as in the case of shredded cheese.

Budget Gourmet's ovenable tray was such a package when it was introduced in the early 1980s. It was different, it made sense, and it saved money at a time when consumers were becoming upset about the high cost of frozen entrees and dinners, blaming overpackaging for a big chunk of the perceived high price. Times and fashions change, in packages as well as foods. A package that is really right for the times is an integral part of the product's success—and may, in fact, be a key reason for its success.

• New packaging and technology created an exciting new category in fresh produce. Packaging for prepackaged salads includes aqua packs; ziplock bags; and modified, atmosphere films that are breathable, patented and used to extend shelf life. As advantages continue to be made in these MAP films, items like sliced apples will be able to join the greens and vegetables without turning brown.

Packaged salads and precut vegetables have made a big difference to hurried shoppers and illiterate cooks who want to eat more fresh vegetables but have felt that too much time, know-how, and effort were required. The new transparent packages changed their perceptions. Quality-wise, packaged salads are rather like the little girl with the curl on her forehead: When they are good, they are very good and when they are bad, they are horrid. Packaged carrots can taste like bitter lemons even before the packages start to swell. Packaged salads can turn to slimy mush when they lose their freshness or their cool. But the convenience of packaged precuts and salads answered real needs. There is a large market of consumers who want to eat more healthfully and more naturally but don't have the will, the discipline, the energy or the time to fix a salad as often as they think they should. In addition, although millions of consumers like salad, often they hate to wash lettuce and go through the time and trash involved in salad fixing. Consumers in both of these groups find the packaged salads a real answer to their needs, even if they have to toss the unused portion the very next day. And even if they perceive it as expensive or have occasionally had the experience of finding the "bagged salad brown and nasty" when they opened it at home:

"It's the best product in years, at least when it's fresh and crisp and good tasting."

"Ready made salads in ziplock bags—real values that fill a real need."

"Von's Ready Pak Salad Mix is the best value we've had this year."

"Dole packaged salads are nice, fresh, clean, reasonable."

- Improved packaging has been a major contributor to the new image of store brands. In fact, improved packaging has generated a whole new breed of shoppers who swing back and forth between name and store brands based in part on whether or not they like the way the store brand choice is packaged. Swing shoppers appreciate good packaging and look for store brand packages that are user-friendly, labels that are readable, and graphics that are designed to clearly signal both the store's mark and the comparable name-brand product. Upgraded graphics and convenience features offer quality cues that these shoppers are ready to receive, communicating that the store is proud of this product and that there is nothing second rate about it. And some retailers in Europe and the U.S., recognizing that their own brand equity doesn't stop with the product, have begun to outpackage the brand packagers.

Consumer Perceptions of Packaging

Consumers have judged packaging harshly for most of the last twenty years. For example, packaging has taken the heat for deceptive marketing practices like hidden size and weight reductions and exaggerated benefit claims. It's blamed for slack fill, broken nails, consumer-be-damned clo-

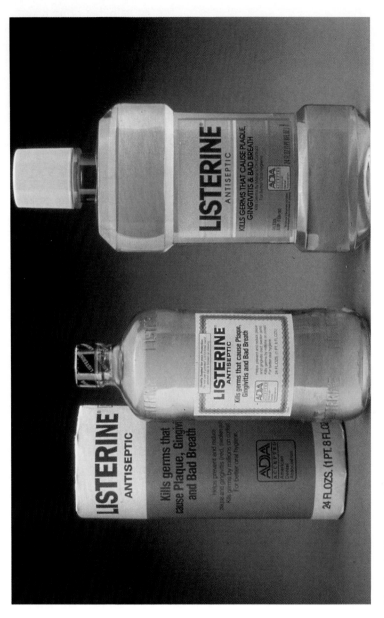

Listerine held on to their package equity long after competitors were in plastic. Customers were attached to the old Listerine package but turned against glass and layered packaging for products they use in the bathroom.

157

sures, nonsensical serving sizes, nonresealability, overloaded trash cans and landfills, and the littered clutter of our throw-away consumer society.

Packaging has been disdained as a step-child of advertising, a necessary evil that is permeated with marketing hype, environmental abuse, and greedy producers. Many consumers have enjoyed venting their frustrations with what they consider bad packaging, poking verbal holes in packages that lag behind their expectations as a way of dealing with the frustrations they have experienced:

- "It took them years to figure out that we wanted packages of hot dog rolls that had the same number of rolls as the hot dog packages had hot dogs."
- "Sanitary napkin packaging makes no sense. The women who need extra long, absorbent pads get fewer pads per box when it is obvious they need more. Okay, raise the price, but put more in a box!"
- "Shaving ounces from the package is the most common way of giving the consumer less for the money."

Praise-worthy packaging gets less fanfare (and much less media attention) than popular advertising, but it does get talked about. It is triggered by packages that work significantly better, deliver more value, are fun to use, are more environmentally friendly, or signal a promotional or an especially consumer-friendly, office-friendly, kid-friendly or driver-friendly reason for buying.

The Total Package, by *Philadelphia Inquirer* reporter Thomas Hines, makes the common mistake of repeating the cardinal rule that people in general think about the package only when it causes them problems. We wonder where Mr. Hines has been for the last five years. Today's consumers are not only capable of talking about packaging they like, but prone to do it. Really good packaging sells product not only off the shelf, but over that greatest of all advertising vehicles, the grapevine. Today's consumers delight in packages that add value and work well. They have had so many packaging frustrations and disappointments that they are surprised and pleased with packaging that catches up with their expectations:

- "Dole reseals in box and in bag; Sun-Maid does not."
- "Really like the Wondra flour package."
- "I like the way Procter and Gamble has a plastic container with lid to refill with powder Tide."

Consumers talk about reseals that make a freshness difference; seals that are suddenly easier to open; designs they find attractive; pull tabs they can grip; shapes and sizes that are easy for them or their parents or kids to hold or carry; dispensers that are easy to use; zippers they can easily zip.

Faced with a load of work and lack of time, today's consumers respond with strong interest to functional as well as aesthetic packaging choices. In some categories, there are more clearcut choices between packaging types than between product alternatives. The packaging choices deliver a wide assortment of benefits. Some packages enable consumers to dispense and use the products they contain in time- and work-saving ways. Some are lunchbox-ready; others go right into the microwave, then to the table. Many are easy to open or prevent messes or provide decorative touches or appeal to environmental sensitivities. . . . Some offer large but portable quantities at significant savings. Many "work" with the product to actually do the task, others work with one hand, or connect with an energy source. Some are entertaining while others showcase the product.

Consumers of the 1990s look for packaging that's easy—easy to open, close, use, store, and dispose of or refill. "I really like resealable food packages." "Anything heavy that doesn't offer refills is inconsiderate of customers, fills more trash cans, and creates more waste." They expect packaging to work without leaking, breaking or collapsing. They want it to be economical and, if possible, refillable as well. At the same time, they want it to be attractive—even beautiful. "Zesty Crackers/too plain." "Store brands/not attractive."

Resealability—especially, but not only as in zipper closures—tops recyclability and environmental friendliness as a reason for best package nominations. Zipper resealables are praised for their convenience, easy storage, and even recyclability. Some consumers went beyond zipper closures—which still require some work—to self-seal packages that close themselves—like the irons that turn themselves off after a few minutes of rest.

Packages that give rise to broad complaints are hard to open or reclose—cereal is the major offender. Some are just impossible to open—especially true of shrink wraps and inner seals. Others (ice cream) come apart or collapse during the use period.

The dramatic growth of packaging choices makes consumers more aware—and more receptive to packaging alternatives. Most of today's consumers are familiar with convenience alternatives, tampering scares, environmental alarms, and the emergence of plastic and other new materials and technologies. Personal experience with packaging alternatives in combination with intermittent headlines triggered by product-tampering tragedies and overflowing landfills and trash barges have given consumers a rich set of "experiences." These experiences have moved basic consumer perceptions of packaging:

- from negative to mixed
- from simple to complex
- from hostile to receptive

As Consumers Approach the Millennium

Time and Money Pressure

Working consumers have less leisure time and less shopping time. Many have less money to spend—many others feel less free to spend what they earn because their future is insecure. They work harder to hold jobs, make ends meet, and keep up with a variety of activities. Additional time pressures come from having more interesting things to do at home—net surfing as well as channel surfing. Getting hooked on computers or the Internet uses up as much time as getting hooked on television.

Many of those who are making money have less time to spend it and many of those with time to spend it say that they really don't have money to spend beyond necessities. Many downsized consumers have found new jobs. Some have created home-based businesses or consultancies. But few are making as much money as they did before downsizing. One shopper explained it this way: "People who have jobs are saving their money in case of a layoff—only buying what they NEED versus what they want. Others can only afford necessities."

Job Fragmentation

More than seventy-five percent of the consumers we surveyed at the end of 1995 knew someone who had lost a job through downsizing during the last few years. The high percentage cuts across all demographic groups, but is higher among consumers under thirty-one and over fifty than among the thirty-to-fifty group. Besides knowing someone else, thirty-one percent had had a downsizing experience in their own household during the last three years.

Trust Shift

More consumers trust themselves while fewer trust institutions. The erosion of jobs and the perceived deterioration of health care in the United States have been accompanied by an erosion of the feeling of being safe—making consumers feel more and more responsible for themselves and their own health and safety.

Technology Explosion

Technology is recognized as a change-maker which benefits those who learn to use it. Keeping up with it has replaced "keeping up with the Joneses" as a source of motivation and stress.

Ethics Imperative (Global, Medical, Personal, Vegetarianism)

More consumers are aware of grey areas and seek information about what's right for their bodies, their families, and the world they inhabit.

Disruption Awareness (Divorce, Disaster, Chaos)

The sense that we must continue to learn to cope with dramatic change has grown dramatically.

Tradition Bending (Clinging)

In the presence of dramatic change, traditions are recognized as valuable and important. At the same time, more consumers are ready to change the way they follow traditions, making it easier to turn to prepared foods for Thanksgiving and Christmas dinners, for example.

New Consumerism

The new consumerism rests on the availability of full information coupled with less time to shop. With the information available, consumers understand that they must rely on themselves for protection and that being a good consumer takes information plus time and energy, both of which may be in short supply.

Fun and Enjoyment

Consumers have decided that it's important to enjoy all aspects of life when they can. They seek and find more sensation, as evidenced by the tremendous growth of big flavors and spicy foods. Long-necked beers and the shapely and nostalgic Coca-Cola bottle capture the spirit and benefit from this major strategic opportunity for packaging.

The Awakening of Packaging Consciousness

The great consumer awakening for packaging began with the Tylenol tampering in 1982. As a result, consumers were given a "seals" and "closures" language that quickly drove them to separate packages from the products they contained. But even before the Tylenol tampering, consumers were getting smarter. Fueled by various perceptions, their purchasing decisions were becoming more complex. These perceptions run the gamut of praise and criticism, reflecting trends, media coverage, personal experiences, lifestyle, and increased expectations for convenience and performance.

Packaging perceptions overlap product perceptions because the package actually interacts with the consumer during purchase, use, and disposal of the product. If interact sounds like an exaggeration of what is an inanimate object, watch a consumer struggling with a jar of sauce, a bottle of ketchup or the liner of a cereal box. Besides shaping consumers' attitudes, these interactions shape perceptions of quality, value, and convenience. They play a role in who, what and how much consumers trust. For many of today's consumers, packaging plays a larger role than advertising in forming and shaping their product images, brand trust, and purchasing decisions.

Here is a cross section of the packaging attributes, features and benefits that come up repeatedly in focus groups and surveys because they matter to today's consumers. All of them offer opportunities to today's marketers.

- aesthetics
- cleanliness
- closures—easy access, opening, closing
- closures—effective dispensing
- convenience in purchase, use and disposal
- cost
- disposability
- ease of use including hand fit
- environment-friendly materials
- freshness dating
- fill (versus settled contents)
- fun to use
- health implications
- handles for carrying and handling
- less (less packaging is better)
- neck bands that add a clean, safe look
- nostalgia
- openability/nail safety
- performance
- portability (and driveability)
- pour spouts
- pull tabs (that work)
- purity/wholesomeness
- quality assurance

- readability and easy comprehension
- recognition (brand/product/flavor identification)
- resealability
- recyclability
- safety
- shelf life at home
- space saving
- sparkle
- tamper evidence
- trust
- transparency
- trigger sprays
- value
- visual excitement

When Perceptions Change

Consumers' perceptions of packaging change in a positive direction whenever their concerns or frustrations are resolved with an improvement that is meaningful to them. Perceptions also change along with changing need states and vantage points. For example, the same consumers see packages differently when they are preparing for a trip than when they are routinely shopping.

Interest in specific attributes and benefits varies with use occasion, product category, budget considerations, lifestyle changes, time-pressure changes, media reports, and the grapevine. Perceptions change every time a really new package is introduced in a category that consumers perceive as important to them or their household or office. New packages create new possibilities and new expectations. In addition to all the discrete changes that take place with shifts of need and perspective, gradual changes take place in response to media reports and advertising and public relations campaigns that are designed to modify perceptions (e.g., to make plastic more acceptable).

Categories where packaging changes have made a significant difference in consumer perceptions in recent years include:

- auto supplies
- bottled water
- brown bag/lunchbox products
- detergents
- fast food

- frozen entrees and dinners
- ice cream/frozen yogurt
- laundry and cleaning products
- juices and drinks
- motor oil
- mouthwash
- meal kits for refrigerated and shelf-stable pizza
- fajitas, lunches, dinners, salads and stir fries
- microwave popcorn
- priority mail envelopes
- produce (precut vegetables and salads)
- puddings
- raisins
- sauces
- shredded cheese
- snacks
- soap
- softdrinks
- toiletries/personal care
- toothpaste
- Tylenol—first with tamper-evident seals and capsules; second with easy-off caps
- unit-dose medications

Significant improvements in all of these categories have softened consumers' negative attitudes toward packaging as a whole and dramatically improved their perception of packaging for specific categories and brands. One example: "Resealable tortilla packages are wonderful." Developments like this that dramatically resolve a frustration can rapidly change perceptions. Sometimes the packaging assortments change almost as rapidly as the perceptions, and a whole category moves from cans to jars or from glass to plastic. Some changes are so rapid that assumptions of how consumers will see a packaging innovation are out of date if they are based on another product category, a different consumer segment, or last year's perceptions. Even performance perceptions change as the novelty of an improvement is replaced by new experience. Cans are still around because they perform extremely well. They are perceived as economical, compact, reliable, and environmentally friendly. Plastics have become the material of choice for more and more packages because many plastic packages perform well. Glass continues to impart a quality image because nothing matches its sparkle or its tactile and taste performance—at least not yet.

Responding to Negative Perceptions

McDonald's conversion from styrofoam to kraft paper packaging in all its U.S. stores is an example both of the impact that negative perceptions can have on a company and the power to turn the negative into a positive with a direct response. McDonald's use of styrofoam generated widespread criticism for being environmentally unfriendly. In response, the company swiftly and dramatically moved to packaging that was perceived to be more environmentally friendly, even though, in the opinion of many technical experts and environmental engineers, the environment would have suffered less with the foam cups and cartons than it will with the newer kraft-colored paper. Public perception dictated the change to seemingly old-fashioned packaging materials that look eco-friendly and assure most customers that McDonald's is doing the right thing. McDonald's "doing the right thing" strategy goes well beyond packaging: In the last few years the company also introduced a low-fat burger and switched to healthier-sounding cooking oils for their fries. Although individually none of these changes is reason for going to McDonald's, collectively, they give the impression of a socially and environmentally responsible company. For packagers who are dependent on public good will, they are a model of responsiveness.

Attitudes, Perceptions and Feelings

To most consumers environmentally reasonable packaging is regarded as the opposite of costly overpackaging. It is optimum, economical and smart.

- "Less packaging and lower prices. Yes!"
- "Absolutely less is better and cheaper. All the plastic bubbles and multiple layers of packaging should be eliminated."

The *less* that consumers want is mostly a matter of what they perceive as excess—the layers, ounces, and inches of packaging and package-size in relation to product-size. Many consumers strongly support refills, the idea being that refills make sense and are what companies and consumers ought to be doing, even though the reality and their own purchase behavior often fail to support their idea. One consumer writes "Bravo to all the detergent companies that sell refill laundry detergent for less money and force us to reuse our containers." That comment embodies the sentiments of many consumers:

- It hails corporate/brand leadership and social responsibility.
- It acknowledges the fact that refills must cost less in order to work.
- It encourages and rewards consumers for being good environmental citizens.

The instant credibility of the "less packaging" claim is a real winner—in special contrast to the hard-to-prove biodegradability and recyclability claims that have been turning consumers on and off in recent years. The "less packaging" claim is obviously and instantly true. Shoppers see this and feel reason to applaud.

Many of the perceptions are powered by strong feelings:

- "Almost every food package starts out with a big lie called 'serving size.'"
- "Plastic wraps on meat packages say PULL HERE TO OPEN but you can't—there isn't enough space to get a grip to pull; sometimes older hands can't pull hard enough."
- "Excessive packaging that violates the environment bothers me a lot."
- "I'm very disturbed by packaging materials that get destroyed in the process of opening or using a package that calls for continued use."
- "Mail-order packaging that is hard to open or too large for the contents is quite irritating."
- "I wish manufacturers would put more food in resealable bags, (e.g., cereal, rice, noodles, flour, sugar, etc.)."
- "Packages that cause a struggle or a hassle should be changed."
- "Single-serving items that are hard to open are often exasperating."
- "I really don't like packages that are bigger than they need to be, especially cereal and crackers."
- "Pull tabs on detergent boxes simply don't work."
- "Cereal packages should be improved because of all the product they waste just because they don't have pouring spouts."
- "Bread without open dating is stupid."
- "Cereal and crackers, especially Nabisco's, that can't be resealed tightly are the worst. It's like they want them to spoil faster."
- "I hate sealed cereal packages inside the boxes that are almost impossible to open, and I don't have crippling arthritis—yet!"

- "Cereal packages are harder to open or I am getting older. You can't get anything open without scissors any more."
- "Anything marked recyclable when there is no place to recycle it bothers me. What a tawdry hoax!"
- "Shrink-wrapping would be okay IF they provided a way to open it without a hatchet and If they included a plastics-recycling ID. I don't want ANY kind of packaging that can't be recycled."
- "Any product that uses two to three layers of wrapping is out of date."
- "Useful information printed on the inside of the package is infuriating."
- "The conversion of glass to plastic now that plastic is somewhat recyclable is socially irresponsible."

A good many of the strong-feelings are product and brand-specific:

- "Post's Blueberry Morning cereal—the liner is impossible to open and the box size is misleading."
- "I'm sure that it's only because they have a near monopoly that Kellogg's still gets away with their very inconvenient cereal packaging."
- "Ibuprofen, Tylenol, etc., capsules—impossible to push out."
- "Efferdent tablets—package is ugly, unhandy, bulky and hard to open."
- "Drano, crystal, impossible to open without breaking nails and losing patience."
- "Fiber Low Fat Snack Bars—teensy bites in big package."
- "Saran from Dow is the worst package."

Every one of the above comments suggests a strategic opportunity.

Environmental Attitudes and Packaging Strategies

Most of today's consumers believe they care about the environment. They've learned to think of environmental protection as concern for Mother Earth that should be everyone's responsibility. They want producers and retailers to care, too.

Few are willing to accept lesser performance or back their caring with extra dollars. But many will support eco-friendlier packaging with their purchasing dollars as long as it delivers performance, value and convenience along with the satisfactions that go with perceptions of doing and supporting the right thing.

400 Shoppers Say That Manufacturers Should Make Their Packages

	Agree Strongly	Agree	Total Agree
More eco-friendly	63%	21%	84%
Easier to reclose	45	25	70
More tamper safe	37	25	62
Easier to open	35	27	62

What consumers really like is environmental responsiveness that is simultaneously consumer friendly, that is, environmentally better and cheaper, environmentally better and easier to carry, environmentally better and healthier too. Environment is an important supporting issue in many purchasing decisions. It tilts rather than triggers purchases, playing an important but secondary role, second to safety, second to cost, second to freshness and quality preservation, even second to convenience and ease of use. Consumers feel very strongly that packaging should be more eco-friendly—even more strongly than they feel it should be easier or more convenient to open and reclose. But the facts of shopping and living put environmental concerns into the second tier, way behind performance. A 1995 mid-western supermarket manager told us that: "Today's shoppers don't even care whether our bags are biodegradable or not. They just want something that won't break open."

In research conducted by Environmental Research Associates in 1994, four in ten adults said they "always" or "usually" look before they purchase products to see if the products carry environmental messages. That percentage appears to be "halo" inflated—most of the shoppers we talk with and/or observe do care and would prefer knowing that much of the packaging they are buying and using is recycled or recyclable and that the bags their stores use are biodegradable. But their priorities in the 1990s are good product performance and minimal hassle. In the case of packaging, this means good performance and ease of use. In the case of bagging groceries, it means getting their groceries home without mishaps. They are welcoming double grocery bagging that uses kraft paper bags for shape and structure inside of a plastic bag, which provides both waterproofing and handles. The paper-versus-plastic battle of the bags at checkouts seems long forgotten.

Today's practical shoppers are also casting doubt on recent source-reduction efforts—especially those involving food. They want packages to be eco-friendly. They want to feel good about them. But first and foremost they want performance.

Excerpt from a 1995 frozen entree focus group:

- "The extra layer is important on frozen products."

- "I've never had freezer burn on any article that I bought with a sealed film across the top of the tray and the tray inside a box."
- "I prefer those packages because when I do buy the Budget Gourmet-types without a seal, I have ice in there when I open them up."
- "That clear plastic keeps out the ice."
- "It seals it in and I have never had freezer burn in a product with that seal."
- "The moisture in the product can end up as pieces of ice, but it is less likely here (in film-sealed plastic tray) because of the other coat."
- "They've come a long way—plastic is safe now."
- "Besides, in the township in which I live, you can recycle plastic containers that have the one or the two in the triangle—most food products come in a package that have a one or a two. The paper is going to go in the garbage and wind up in a landfill somewhere."

Protecting product quality and enhancing taste by avoiding ice crystals, freezer burn, and the miserable mouth feel that goes with imperfectly handled frozens seems much more important to frequent frozen food users than the benefits of removing a packaging layer. To many users, enhancing the eating experience with food trays that look kitcheny seems more important than eating out of a cardboard box to save a dime or land-fill space.

Purchase Decisions versus Environment Concerns

Most shoppers continue to base most purchase decisions on their product or brand preferences coupled with perceptions of price and convenience—even if they say they care a lot about the environmental factor. In spite of this marketplace behavior reality, packagers who dismiss environmental concerns and pay attention only to purchase behavior do so at high risk.

Package and brand strategists who can think beyond the next quarter should think about the amount of purchase influence that can be expected in the future from choosing a more or less environmentally friendly option. Not every company is as dependent on perceptions of good corporate citizenship and "right" behavior as McDonald's, whose packaging move toward kraft-looking paper was discussed at the beginning of this chapter. But almost any packaged product is vulnerable to a

competitive product that suddenly appears in an environmentally friendlier package that is otherwise equal. And the less environmentally friendly
the established package appears to be, the more consumers are likely to
switch to the suddenly friendlier competitor(s).

The Case for Comprehensive Packaging Research

Some consumers don't think that food tastes good when it rubs up
against paper. Others don't think food is adequately protected when it's
packaged in a single layer of paperboard. A twenty-five-year old says: "I
don't feel like eating my meal in a cardboard box." A thirty-four-year-old
shopper sees the worst package in the supermarket as "Aunt Jemima Pancake Mix—no inside liner for freshness, just a cardboard box." Many consumers think that plastic liners inside of paperboard boxes are antiquated:
"Cereal—there's got to be a better way than wasting a whole box and an
inner liner of plastic." Others believe that packaging improvements would
raise "already horrible" cereal prices even higher.

These conflicting perceptions present great opportunities and a mine-
field of danger. Comprehensive research and analysis is sorely needed in
this area. The pressure for timesaving in the product-development process
has not only curtailed strategy-oriented research, it has curtailed the development of improved packages. Time and cost pressures have been used to
justify the elimination of home placement tests in favor of faster mall in-
tercept research (intercepting mall shoppers who meet product use and/or
demographic criteria to get their reaction to the "improved" package).
This sidesteps real home-based and on-the-go experience with the package. It misses the long-term frustrations and performance perceptions that
not only extend beyond the initial point of purchase but continue to be
important throughout and beyond the life of the product.

Most of the package market research that does get funded focuses on
the prepurchase, shelf impact of the current or proposed packages. Such
research addresses questions that are significant to getting noticed and
generating product trial. It's not surprising that designs and attributes that
tilt initial purchase get marketing attention and research dollars.

Failure rates for new products run as high as eighty percent in the
United States, but few packaging studies look at the role of packaging in
repeat purchase decisions. This disregard for the complex role of packaging in new product failures continues for three reasons:

1. The kinds of packaging questions that are most frequently asked encourage consumers to dismiss the importance of packaging in their re-

sponses. The 1994 *Packaging Magazine* study reported that only 1.2% of its respondents agree that "a new package will make them buy a different brand." The actual question asked generated the response obtained. Think about it: How many consumers do you think would say or even admit to themselves that the package is more important than the product? How many interpreted the "new package" as a new color, new design, new twist, or new amount? Asking the question a different way would have produced a different answer. For example, how many respondents do you think would have said that a new package would be a reason to switch brands if the new package was described as a "more resealable package at the same price"? In our experience at The Consumer Network, the response rate would have been more like 71% than 1%.

2. Research projects are more likely to get funded when a positive payoff is expected. Showing which color, copy point, attribute, closure, spout, or design generates the most interest or the highest purchase intentions promises a high rate of return on research investments.
3. Packaging isn't easy to research because consumers don't always separate it from the product.

Users' experience of the package is closely related to their attitudes about and trust in the product and the brand. But corporate and brand attention to the part that package satisfaction plays in product satisfaction receives more lip service than actual research attention. Marketers tend to recognize the importance of packaging on one hand and take it for granted on the other. They are quick to insist that most of their consumers equate their packages with their products. They are aware of negative feelings about packaging but unlikely to agree that packages play a pervasive or as persuasive a role as advertising and promotion in maintaining customer trust and loyalty. In an era of spin doctors and widespread lying, negative attitudes about everything get regular doses of media attention. Consumers learn to expect lies, and marketers learn to either dismiss or manage grumbles with complaint-handling systems in order to keep moving ahead. Stories about the moods, outrages, annoyances, resentments, and disappointments of one group or another are as routine as local homicides. Marketers have had to learn how to pay close attention to trends—because they definitely impact the bottom line. In the process, they have learned to look at consumer frustrations as marketing opportunities to be addressed at the convenience of the business. In this climate, trust continues to erode. In response to a survey question about which brands were trusted most, one consumer wrote: "You know, I have come back to this question three times and the answer is no, there is no brand I trust any more."

Longitudinal studies of packaging satisfaction are the exception rather than the rule and are most likely to be undertaken in response to long-range problems like tamperings or the threat of environmental legislation. When home-use studies are done, they are likely to focus on satisfaction with the product and comprehension of the instructions rather than specific satisfaction with the package. The growth of 800 numbers theoretically ensures that brands and companies are going to get earfuls of packaging complaints that arise during the use phase. Some companies believe they don't need to do longitudinal user studies because feedback on home-use problems comes to them via 800 lines.

The relationship between packaging satisfaction and product repurchase sounds academic and is not likely to get brand or corporate funding. One "sad" packaging story in which a packager's failure to conduct longitudinal real-life performance tests with target users resulted in the total collapse of a product that had high consumer interest and high corporate success expectations. The fact that the package looked purse-portable was a major factor in the high consumer interest and initial purchase. The fact that the package leaked all over women's purses after getting pushed around with other purse contents triggered lots of complaints and few repurchases.

That's not an isolated story. Lorna Opatow's chapter is aptly entitled "Getting It Right." Consumer satisfaction with package performance is rarely used as a measure of brand success. Consumer dissatisfaction with packaging performance and credibility has been viewed as a given. Twenty years of frustration with cereal packages hasn't stopped consumers from buying cereal. Frustration with packaging is recognized as an opportunity for marketing after it becomes necessary to change the status quo in order to comply with new regulations or reposition a product or product line to gain or maintain market share.

Where, Oh Where, Is the Packaging Research Foundation?

The Advertising Research Foundation (ARF) has been in business for a long time. It was formed back in 1938 to embrace the needs of four separate constituencies with a common objective: improving advertising productivity. The four constituencies of the ARF are advertisers, advertising agencies, research companies, and the media—and its chairpersons continue to rotate among the four parts of the industry. The ARF is not nearly as strong today as it was in its heyday, in part because advertising itself is not as strong as it once was and also because the pressure on advertising budgets has squeezed research departments out of many ad agencies.

Nevertheless, the ARF has done and still does excellent work. It has helped showcase the role of research in advertising success. It hasn't solved advertising's credibility problem, but it has addressed it from several angles and has a very professional reputation for using comprehensive and sophisticated research among its own members and the business community at large. The ARF doesn't reach a consumer audience or have a consumer profile, but it does have an image of professionalism and credibility within the business community, and it elevates the role of advertising research throughout industry.

Tony Adams, author of our entertaining but provocative chapter on red and white advertising, served as one of the ARF's most popular presidents. Tony's refreshing insights are one reason for discussing the Advertising Research Foundation in a book on packaging. There are at least three others:

1. For packaged-goods marketing, advertising, packaging and research are closely intertwined and almost certainly should be even more intertwined than they are already.

2. Advertising pros are, almost by definition, better communicators than packaging pros. This should be no surprise, since persuasive communications is their primary business. Advertising is also more glamorous than packaging, and has traditionally received more attention and discretionary marketing dollars from brand management and top management. The stars of the packaging industry are great package designers and engineers. However, few designers and engineers hold a candle to advertising experts as communicators. The stars of advertising are creative geniuses who have mastered the art and science of persuasion. No wonder the market power of packaging frequently takes a back seat to the internal sales power of creative advertising!

3. Although there are literally dozens of packaging associations to choose from and to represent various parts of the packaging industry, there is no Packaging Research Foundation counterpart of the Advertising Research Foundation. The annual consumer survey sponsored by the now defunct *Packaging Magazine* took a snapshot in less than twenty-five questions and tended to interpret, soften, and "sell" the results to readers of the magazine in ways that maximized their comfort and downplayed the messages that consumers are sending. It reported that "More than half of the consumers surveyed think foods, cosmetics and pharmaceuticals use too much packaging." Fortunately, the false image of "overpackaged" appliances and electronics is changing for the better.

Astute readers should have been alarmed by reporting that presented image and perception findings as *false images* when measuring images was the mission of the survey! Reporting the results in terms of the percent of consumers who hold the "false" image puts a positive spin on the findings and assures readers they really don't have to worry about them. The same survey reported what consumers rated important about packaging benefits but failed to report what consumers found frustrating, irritating, hassling, or just plain dumb about the packages they were experiencing. The still mythical Packaging Research Foundation would do a much better job.

Conclusion

The improvements in packages and the growing power of packaging aren't happening haphazardly. Consumers are getting more benefits from packaging, they are talking about these benefits in surveys and focus groups and they are using them in their purchasing decisions. The interactive packages on the horizon will deliver even more benefits and call for even more consumer understanding. Progressive firms will increasingly develop proactive packaging strategies that utilize consumer research and take consumer perceptions, trends and needs into account. Better packaging will make better marketing, and vice versa.

The best strategies will stand up against profitability objectives, fluctuating material costs, a changing marketplace, and frequently changing and conflicting consumer perceptions. They will also show understanding of consumers as well as technology, taking into account not only the effectiveness of the package in selling the product on the shelf, but also the power of the package to resell the product during its use. And they will make it possible for the company and/or its brands and divisions to develop and implement packaging changes that differentiate products, satisfy customers, and enhance product and brand values.

A Look Ahead: Packaging Prognostications for the New Millennium

From Herb Meyers

- Talking packages that provide information about the product inside, use instructions or cooking instructions.
- Printed circuits that give a loud noise if pilferage or tampering occurs on the package.
- Printed circuits that can interact with computers or video circuitry to make mandatory copy legible for the elderly—or anyone for that matter—especially in connection with medicinal products.
- Programmed circuitry that interfaces with microwave ovens providing proper food preparation automatically.
- Magnetic UPC codes that read and add up costs automatically when putting the packages on a check-out conveyor. Bills are paid as you exit through a credit card system similar to the automatic cash machines at banks.
- Magnetic UPC codes could also control inventory in the store and restock through an automatic stocking and transport system that moves supplies from warehousing facilities underneath the stores.
- Tablet/caplet/capsule pouches containing the proper dosage of medicine that make a beeping sound at preset times when the medicine should be taken.
- Touchstone system to aid in opening and closing packages and also is childproof.
- Bread bags and inner cereal bags with zippers for easy opening and closing providing improved freshness of the products.

- Cans that open more easily through tear strips or pull handles thereby eliminating the need for struggling with can openers.
- Impossible concepts? Remember Dick Tracy of a few years ago with his radio watch? It's here today.

From Greg Erickson

- Makers of aluminum beverage cans—after getting clobbered by PET in soft drinks and losing out to glass in the super-premium beer market—will rejuvenate their containers with wild shapes, eight-color printing, wide-mouth ends, and even resealability features. Then it will be up to the beverage companies to invest in these attention-grabbing advancements.
- Following "light" beverages, "crystal" beverages, and all-natural "new age" beverages, expensive *indulgent* beverages— milkshake-thick concoctions in relatively small portions formulated for the adult palate—will make their debut as a breakfast "meal," lunch-time dessert, or before-bedtime snack. Spicy flavors, savory flavors, and herbal flavors all might appear along with sweet ones. Packaging will play a crucial role in designating them as something entirely new.
- The composite can will face competition from flexible-film pouches and bags for frozen orange juice concentrate. These bags, superbly decorated, will hang from pegs in supermarket freezer cases. The concentrate will thaw quickly and be dispensed easily from such pouches.
- Too little attention will be paid by product marketers and package designers to the changing needs and wants of the aging U.S. population. Therefore, many otherwise excellent products will fail. Long term, we might even see organized protests against packaging that older consumers find insulting.

From Mona Doyle

- Smart packaging that interacts with consumers, computers and appliances will dramatically escalate consumers' expectations of direct packaging benefits.
- Online shopping will put new demands on packaging and blur the lines between packaging and marketing.
- The corporate profile for packaging and packaging research will equal the advertising and promotion profiles.

- Back and side panels will be as well designed and shopper-friendly as front panels.
- A level of attention and creativity to product-usage communications that parallels the attention and creativity given to image advertising.
- Heatable/reheatable coffee, tea or soup cups that make and keep beverages hot while working or driving.
- UPC codes that tie directly into home and store computers for complete background on the product—from production and marketing to ingredients, interactions and nutritional analysis.
- A proliferation of rational sizes that return a sense of value and trust to packaged products.
- The emergence of fun as an adult marketing tool impacting packaging shapes, colors, and functions.
- The emergence of multifunction packages (e.g., large detergent boxes that can be used as stepladders and footstools).
- Consumer-friendly packages for cereal and maybe for flour and sugar as well.
- Wide use of the internet will facilitate consumer research on new package designs and attributes.
- A Packaging Research Foundation, funded by major packagers, will sponsor innovative and comprehensive research that among other functions, will document the contribution of packaging to product success and failure.